Pre[illegible]

This Study Guide is designed to accompany the textbook [illegible] i-*verse, Third Edition*, by Theodore P. Snow. The chapters in this guide parallel those in the text. Here each chapter contains a summary of the corresponding chapter in the text; a list of words, phrases, and names with which to be familiar; an exercise designed to clarify an important concept related to the subject of the chapter; a multiple–choice self test; answers to selected problems from the text; answers to the self test; and a list of additional reading materials to complement the list at the end of the chapter.

There are certain similarities between learning science and learning a foreign language. Both require that you master vocabulary words, and both require grammar in order to use the vocabulary. The grammar of science is mathematics, and a feeling for what astronomy is all about can only come from experimenting with and understanding this tool.

There are a number of ways in which this Study Guide can help you to learn the vocabulary and tools of astronomy. After you have studied a chapter in the text, turn to the corresponding chapter in the Study Guide and read the summary. Ask yourself whether you understand all the concepts and can elaborate on them. Next, review the list of words and phrases, and return to the text for an explanation of any that you cannot define or explain. Many of these terms will appear repeatedly in later chapters, so it is essential that you learn them.

Astronomy is not just a notebook full of facts, and vocabulary alone is certainly not science. Science is a way of thinking about problem solving. The text includes many examples that illustrate the scientific approach to problem solving, but you never know how well you understand a problem until you try it yourself. To test your understanding, start by doing the problems at the end of each chapter in the text. The answers to several of these are given in the Study Guide, so that you can check yourself. These answers have been entirely rewritten, since all the questions and problems in the third edition of *The Dynamic Universe* are new. Each chapter in this guide contains an exercise to further strengthen your understanding of the concepts of the text. These exercises are not just problems, but are designed to lead you through many of the steps towards a solution. A simple calculator will make your job easier and faster, but is not essential. The exercises are all self–contained.

There is a self test in each chapter, with answers at the end. In reviewing your answers, ask yourself why the wrong answers are wrong, as well as why the correct ones are correct.

At the end of each chapter there is a list of additional readings. The articles were selected because they are relevant to the chapter, are current, appear in books or magazines that are available in most college libraries, and are written at an appropriate level. These reading lists should be a useful tool for anyone wishing to explore a specific topic further, or for the student writing a paper. Several sources, including *Scientific American, Sky and Telescope, Mercury, Astronomy, American Scientist, The Griffith Observer, Physics Today*, and *Smithsonian* magazines, were searched thoroughly for suitable articles, and these reading lists represent a broad selection of astronomy articles published in these magazines over the past several years, with the exceptions that articles thought to be too dated or not strictly relevant to material in the text were omitted. The coverage of *Scientific American* and *Mercury* is complete from mid–1985 back to 1970 (actually 1972 in the case of Mercury, that being the year of its inauguration), and the others are covered from mid–1983 through 1975.

Much of the final preparation of this Study Guide was accomplished by K.S. Bjorkman, who wrote many of the answers to the questions from the text, and who also converted the manuscript to camera-ready format. J.O. Bennett also helped with the answers to the questions from the text. In addition, we are grateful for the editorial guidance of D. Simon and T. O'Dell of the West Publishing Company.

We hope that this Study Guide will be a valuable resource for students getting their first introduction to astronomy, as well as for those who are already somewhat familiar with the subject. We can all appreciate the visual beauty of astronomy; this guide is designed to help the student appreciate the intellectual beauty as well.

Catharine D. Garmany

Theodore P. Snow

May, 1988

Contents

Study Guide for ***The Dynamic Universe, Third Edition***

C.D. Garmany
T.P. Snow

Chapter 1

The Essence of Astronomy

The subject of astronomy encompasses the entire universe, and today our science is more properly called astrophysics, as the laws of physics are applied to the heavens. But astronomy is also one of the oldest areas of human inquiry, the basis for all scientific study. In ancient times the interpretation of what was seen in the sky was sometimes based on myth, but starting in the seventeenth century astronomers have applied laws of physics to phenomena observed in the heavens.

The immense distances between objects in the sky are often difficult to comprehend, so it is helpful to reduce things to a smaller scale, as in a model train layout. If we thus reduce the earth to the size of a basketball, then the rest of the solar system can be envisioned as a handful of balls of various sizes, spread over a circle some two hundred miles in diameter. On this scale the nearest star is about ten times more distant than than actual distance from the earth to the moon, so a new scale is needed if we are to appreciate the extent of the Milky Way galaxy. If we further reduce the solar system so that it is a circle 8 centimeters across, then the nearest star is 300 meters away, and our galaxy is about the size of the United States. Even further reduction is necessary to create meaningful scale models of the galaxies in our local cluster.

Distances in astronomy are sometimes measured in terms of the distance travelled by light in a given period of time; thus we speak in terms of light years. When we examine distant galaxies, which are seen as far as millions or even billions of light years away, we are seeing them as they were when the light left them, long ago. Our study of astronomy will start close to home and eventually take us to the edge of the observable universe.

Words and Phrases to Understand

- astrology
- astronomical unit
- astronomy
- astrophysics
- celestial sphere
- comets
- ecliptic
- light year
- meteor
- Milky Way galaxy
- planets
- revolution
- rotation

Exercise: What Do Scientists Do?

It is sometimes said that the best scientist is the one who asks the right questions. As Stephen Jay Gould writes in *Hen's Teeth and Horse's Toes* (Norton & Co.: New York 1984), the prerequisite for great science is presenting a question in a new and different way. While very few people have the insight to ask questions which will change the direction of scientific research, anyone who studies science soon realizes that science is a method of dealing with problems, not a collection of facts. Astronomy is no exception. Although it differs from the laboratory sciences in that few aspects of astronomy can be directly examined, it shares with other branches of science a method in which the development of general ideas leads to specific theories which can be tested and which must comply with the known laws of physics. Since astronomy covers such vast territory, organization of the available data is especially important, and it is often this organization which leads someone to ask a new and crucial question.

Go out on a clear night and look at the sky for awhile. After you come in, make a list of the questions that came to mind while you were outside. These would be questions about the things you saw, perhaps about their sizes and distances, their motions, their compositions, or their relationships to each other. Having the same observational tools (your eyes) as our ancient ancestors, you will be able to ask the same questions they did. During your study of astronomy, you will see how the ancients eventually found answers to their questions, and you will see how many of your questions are answered. As your knowledge increases, you will be able to ask more sophisticated questions, and will come to understand some of the important ones beings asked today by astronomers at the forefront of research.

Self Test

1. Today, astronomy is synonymous with
   a. astrology
   b. astrophysics
   c. space exploration
   d. geophysics.

2. A moonless night sky is more likely to occur
   a. when it is heavily overcast
   b. when the moon is full
   c. when the moon is a thin crescent
   d. in the winter.

3. The particular stars that can be seen on any given night depend on
   a. the brightness of the moon
   b. the season of the year
   c. the time of night
   d. our geographic latitude
   e. all of the above.

4. The earth revolves in 24 hours and rotates about the sun in 365 days.
   a. true
   b. no; the terms "revolve" and "rotate" should be interchanged
   c. no; the earth rotates about the sun in one year
   d. false; the sun rotates about the earth.

5. An astronomical unit is
   a. a distance equal to the earth–moon distance
   b. a distance equal to the sun–earth distance
   c. the distance light travels in one year
   d. one course in elementary astronomy.

6. How many degrees are there in a circle?
   a. 90
   b. 180
   c. 360
   d. 365.

7. A light year is
   a. a way to measure time
   b. a way to measure distance
   c. the velocity of light
   d. the distance from the earth to the sun.

8. The path on which the sun appears to move among the stars around the sky each year is called
   a. the Milky Way
   b. the zodiac
   c. the equator
   d. the meridian
   e. the equinox.

9. As you travel from the pole to the equator
   a. the setting stars make a greater angle with the horizon
   b. the rising stars make a smaller angle with the horizon
   c. the setting stars make the same angle with the horizon
   d. there are more circumpolar stars.

10. We can distinguish planets from stars because
    a. planets do not twinkle and stars do
    b. planets move with respect to the stars over the course of weeks
    c. planets are brighter than most stars
    d. all of the above.

ANSWERS TO SELECTED REVIEW QUESTIONS

Thought Questions

1. *Scientific theory* is the process by which one attempts to explain and understand the workings of the universe in terms of physical laws. It makes predictions which can be tested by experiment. *Speculation* does not necessarily rely on physical laws, nor does it necessarily make predictions which can be tested. *Dogma* requires only belief, and is generally considered a matter of faith rather than a rational explanation.

4. Since the path of a planet across the sky indicates the plane of its orbit about the sun, the fact that all the planets appear to travel along the same strip across the sky tells us that the orbits of all the planets must lie in almost the same plane.

6. Many objects in the northern hemisphere sky are not visible from the southern hemisphere because they are below the horizon. Similarly, northern hemisphere observers cannot see objects in the southern hemisphere which are below the horizon. Therefore, in order to observe all objects in the sky, observatories must be located in both hemispheres. At the equator (latitude 0°) the entire sky can be observed at some time during the year, although objects very far north or south will be extremely close to the horizon, and hence cannot be observed very well.

Problems

1. The conversions to scientific notation are:

$$0.0045 = 4.5 \times 10^{-3}$$
$$12,400,000,000 = 1.24 \times 10^{10}$$
$$120 = 1.2 \times 10^{2}$$
$$0.0000000000067 = 6.7 \times 10^{-12}$$
$$187,000,000,000,000,000 = 1.87 \times 10^{17}$$

3. To convert from degrees (°) to arcseconds (″), we must use the conversions from degrees to arcminutes (′) and from arcminutes to arcseconds:

$$360° = 360° \times \frac{60'}{1°} \times \frac{60''}{1'} = 1,296,000''$$

5. The speed of light is about 3×10^5 $km\ s^{-1}$, so the light travel times from the sun to the planets requested are as follows:

Planet	*Distance from Sun*	*Light travel time*
Earth	1.496×10^8 km	8.3 min
Jupiter	7.78×10^8 km	43.2 min
Pluto	5.90×10^9 km	5.5 hr

6. If galaxies formed 10 billion years ago, then we must look back to a time earlier than that, which means we must look out to a distance of more than 10 billion light years away.

Answers to Self–Test

1–b, 2–c, 3–e, 4–b, 5–b, 6–c, 7–b, 8–b, 9–a, 10–d.

Chapter 2

Cycles and Seasons: Motions in the Sky

It took mankind thousands of years to realize that the motions seen in the heavens can be simply understood when it is recognized that we view everything from a moving platform, the earth. Once mankind accepted the idea that the earth orbited the sun once a year and that the other planets orbited the sun with periods that are determined by their distance from the sun, the mystery of celestial motion was solved.

The largest apparent motion is due to the rotation of the earth every 24 hours and is obvious in less than an hour's observing. This causes the sun, moon, planets and stars to rise in the east and set in the west. It also defines our timekeeping system. The second obvious motion in the sky is that of the moon, which orbits the earth every 27 days. Its motion against background stars is noticeable over a period of a few hours, and its changing phases mark its progress from night to night. The orbital motion of the earth around the sun causes the sun to appear to move eastward with respect to the very distant stars. This motion is noticeable after about a week of observing. The stars appear to set about 4 minutes earlier each night, and gradually the appearance of the evening sky changes, so we speak of the "autumn" sky or the "summer" sky.

The planets, in orbit around the sun, also change their positions with respect to the background stars, reflecting both their own and the earth's motion. Mercury and Venus, closer to the sun than earth, swing from one side of the sun to the other and thus are seen only near sunrise or sunset. The outer planets are not so hampered, and move gradually eastward, their motion depending on the sizes of their orbits. Only when the earth passes them do they appear to reverse direction and to briefly move westward.

The orbits of the moon and all the planets lie almost in the ecliptic, so most of the motions we are describing take place along a narrow path in the sky called the zodiac. The earth's axis of rotation is tilted by $23\frac{1}{2}^\circ$ from the ecliptic, and this is the cause of our seasons. If the orientation of the orbital planes were identical, we would see monthly lunar eclipses when the moon entered the shadow of the earth, and solar eclipses when the moon's shadow fell on some position of the earth. Instead, eclipses can only take place when the moon passes through the earth's orbital plane.

Words And Phrases

angular diameter
celestial equator
conjunction
constellations
declination
deferent
diurnal motion
eclipses, solar and lunar
epicycle
equinoxes, vernal and autumnal
heliocentric
meridian
Milky Way
opposition
orbital motion
phases of the moon
planets
precession
quarter moon, first and third
right ascension
retrograde motion
rotation
sidereal period
solar day
solstice, summer and winter
stellar parallax
synodic period
superior planets
zenith
zodiac

Exercise: Spatial Orientation

In astronomy it is very important to be able to visualize things spatially; that is, to mentally view something from different directions when it is impossible to actually do so. The following exercise is designed to help you develop these skills.

We start with our direct observations: on some particular clear night in spring we go out to stargaze. We know the cardinal directions: north, south, east and west. The first quarter moon is on our meridian just as the sun sets, and someone asks what the moon looks like at that moment to people several time zones to the east (for example, if we are in Colorado, how does the moon appear to a New Yorker?) To answer this, let us imagine that we could view the earth-moon system from a point in space far above the north pole:

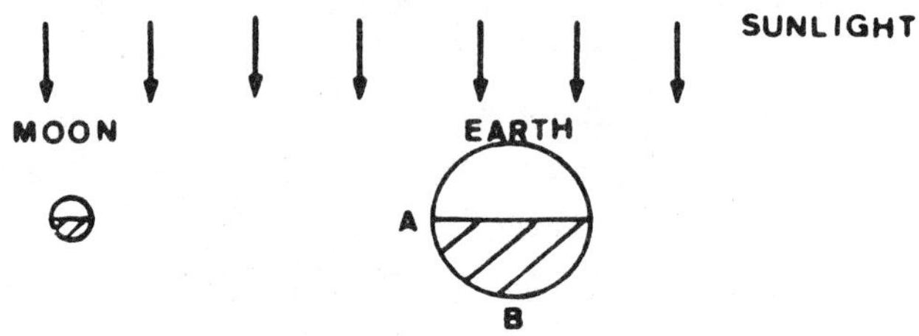

Our location at sunset is at point A. About 6 hours later the earth's rotation has carried us to point B, but the moon's motion is negligible. Therefore the moon is setting in the west. Keep in mind that the direction "east" is simply the direction in which the earth is rotating, and "west" is the opposite direction.

Now ask yourself the following questions: what is the phase of the moon when it is overhead at midnight? During what hours can you see the third quarter moon? In general, what fraction of the time is the moon in the daytime sky? What does the moon look like to people several time zones to the east, our original question?

Now let's expand the problem. The time is about 2 hours after sunset. Close to the western horizon we see Venus. Mars is on our meridian, and Jupiter has just risen. What would the solar system look like if we could view it from far above the ecliptic plane? We would see the following, with the orbital paths of the planets partially drawn for clarity.

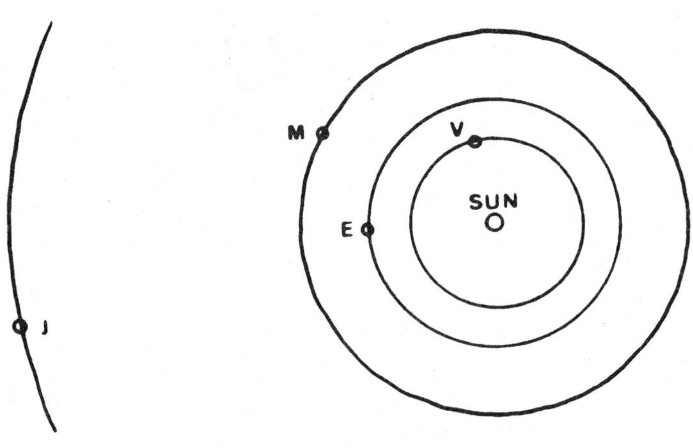

Try the following problems in the same way (your answers will not be exact, but that is not important):

a) You get up an hour before sunrise. The moon is just setting. Just above the eastern horizon you identify Mercury, and higher in the sky, Venus. Jupiter is about to set and Saturn is on your meridian. Draw the planets as they would appear from above the ecliptic plane.

b) With some information about the rate at which the planets move, you can predict what the sky will look like at some time in the future. Skip ahead in the text and find out how much of its orbit each planet will complete in three months. Now redraw your figure from above, not forgetting that the earth will move also! What will the morning sky look like now?

Finally, let's turn the problem around. On March 10, 1982, the configuration of the planets as seen from above the solar system was as shown on this page. This is as close to mutual conjunction as the planets will get for another 179 years, and it gained attention because of a book entitled "The Jupiter Effect" written in 1974. The authors, John Gribbin and Stephen Plagemon, predicted that serious disasters in the form of earthquakes would occur on this date. Scientists studying their theory found no basis for it, and even the authors revised their views after the book's publication, but many people heard about it and were apprehensive. Nothing catastrophic occurred on March 10, but many people observed the planetary configuration. From the figure, describe what the planets looked like as seen from earth and when you could have seen them best.

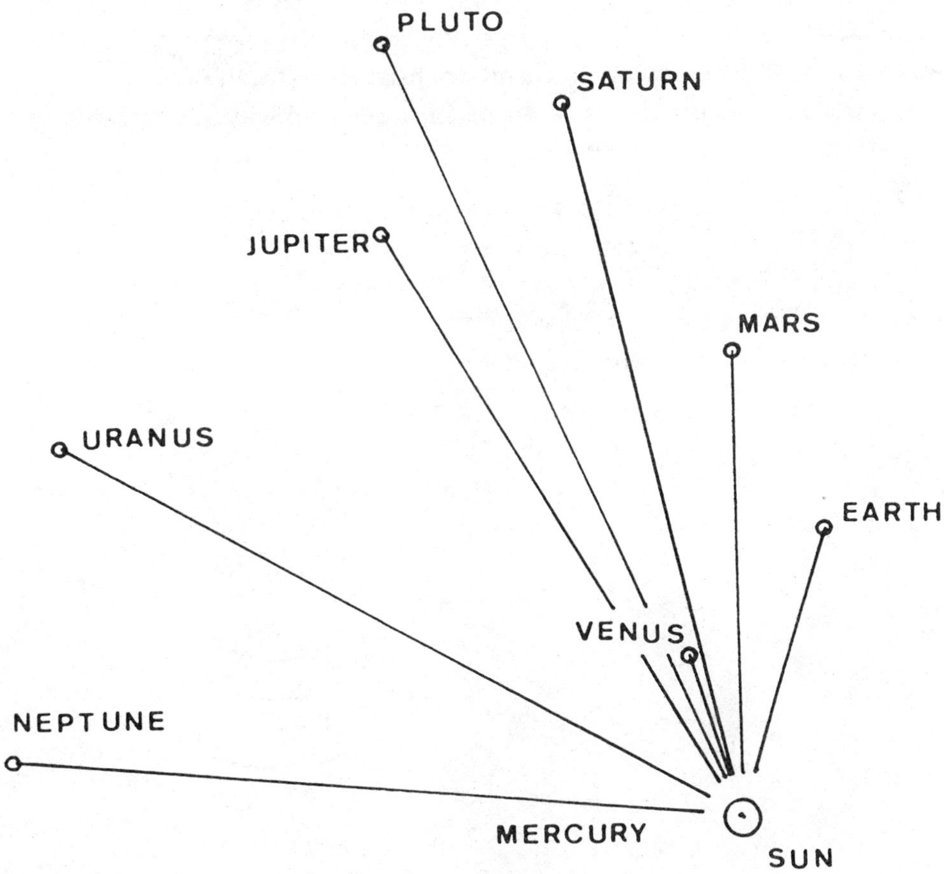

Self Test

1. At the summer solstice (June 21) the declination of the sun is
   a. $+23\frac{1}{2}°$
   b. $-23\frac{1}{2}°$
   c. 0°
   d. different every year.

2. If the moon completely covers the sun as seen by an observer on earth, it is called
   a. a total lunar eclipse
   b. a total solar eclipse
   c. a partial lunar eclipse
   d. a partial solar eclipse
   e. an annular eclipse.

3. The first quarter moon would be expected to
   a. rise at noon
   b. set at noon
   c. rise at 6 a.m.
   d. set within 3 hours after sunset.

4. We experience seasonal variation of temperatures on the earth primarily because
   a. the orbit around the Sun is not circular, so the amount of heat we receive varies
   b. the axis of rotation is tilted with respect to the orbit
   c. the heat from the sun varies in a regular way due to sunspots and solar flares.

5. The motion of the planets with respect to the stars
   a. is visible after a few hours of observing
   b. is visible after a few days of observing
   c. is different for each planet
   d. is not visible to the naked eye.

6. An observer on the earth's equator will see:
   a. stars move parallel to the horizon as the earth turns
   b. the entire celestial sphere in the course of a year
   c. the north celestial pole but not the south pole
   d. the sun pass through the zenith every day.

7. The effect of precession is to
   a. gradually change the angle or inclination of the earth's axis
   b. gradually change the shape and appearance of the constellations
   c. gradually change the constellations that define the zodiac
   d. gradually shift the intersection of the equator and ecliptic along the ecliptic.

8. The first to assert that the earth was round was
   a. Pythagoras
   b. Aristotle
   c. Plato
   d. Ptolemy.

2. Over the span of one year, this difference amounts to about 24 hours. This means that at the end of one year, the sidereal day and solar day will be synchronized once again, and the stars overhead at midnight will be the same ones that were overhead at midnight the previous year.

4. The Magellanic Clouds, at a declination of $-70°$, as seen from a latitude of $20°$ would have an apparent altitude in the sky of $0°$, and so would be just on the horizon from the observatory at Mauna Kea. This nearness to the horizon would make them very difficult to see, although not impossible (since Mauna Kea is at a very high altitude).

6. At the winter solstice, the sun has a declination of $-23\frac{1}{2}°$. If you lived at a latitude of $30°$, a declination of $+30°$ would be directly overhead. The sun would therefore be $30 + 23\frac{1}{2} = 53\frac{1}{2}$ degrees away from overhead, toward the south, which would mean that the sun would appear at $90 - 53\frac{1}{2} = 36\frac{1}{2}°$ above the southern horizon. At the summer solstice, the sun would have a declination of $+23\frac{1}{2}°$, and would therefore appear only $6\frac{1}{2}°$ away from the zenith, or $90 - 6\frac{1}{2} = 83\frac{1}{2}°$ above the southern horizon.

8. If one adopted a year of 365 days, then an error of 0.256 days would accumulate over each year. After 4 years, the accumulated error would be $4 \times 0.256 = 1.02$ days. It is to make up for this one day error that we insert an extra day every 4 years, known as a leap year.

10. The equation relating the angular size of an object, θ, to its actual size, s, and its distance, d, is:

$$\theta = \frac{s}{d}$$

.

This means that if the distance were doubled, the angular size would only be half as large. Then total solar eclipses would never occur. All solar eclipses would be annular eclipses. If the distance were halved, the angular size would double. In this case, all solar eclipses would be total, and the effects known as "Bailey's Beads", in which the sun shines through the cratered edge of the moon, would not be observed, nor would many of the flares and prominences on the edge of the sun be observed during totality.

Answers to Self Test

1-a, 2-b, 3-a, 4-b, 5-c, 6-b, 7-d, 8-a

Additional Readings

For this chapter on the nighttime sky, the best supplemental sources of information are those that point out objects and events that may be viewed with little or no special equipment. For this purpose there are several guides and handbooks, such as the Observer's Handbook (published annually by the Royal Astronomical Society of Canada), which contain tables of planetary positions, the motions of the sun and moon, dates of prominent meteor showers, and lists of interesting objects such as the star clusters, nebulae, and galaxies from the Messier catalog.

There are also some periodicals in which astronomical information for the amateur observer is published. Perhaps the most complete of these is the magazine Sky and Telescope, which includes monthly star and planet charts and reminders of specific events, such as meteor showers, planetary configurations, and the appearances of comets. Sky and Telescope also includes other interesting features, such as articles on topics in astronomy, notes on current developments in the field, information on observing techniques for amateurs, and even hints for telescope makers. Other periodicals that provide some of this information include Mercury (published by the Astronomical Society of the Pacific, a unique association of professional and amateur astronomers), the Griffith Observer, Observatory and Astronomy.

Chapter 3

Early Astronomy

From earliest times, mankind has been preoccupied with the sky, and considered it to be highly significant to his every-day affairs. The earliest evidence of this comes from the British Isles, where prehistoric people, as early as 2000 B.C., built Stonehenge and similar structures which show surprisingly advanced astronomical knowledge. The dawn of understanding of the sky occurred in many parts of the world around 500 B.C., however. Although parallel developments took place in Asia, India, the Americas, and the Mediterranean region, it is from the latter region that we have inherited most of our nomenclatures and ideas.

Babylonian astronomy arose around 1700 B.C., and cuneiform tablets from that period show that these people knew a great deal about cyclic motions in the sky. Eventually, by about 200 B.C., it was possible for the people of this region to predict eclipses and planetary motion, and they developed a cosmology to fit their world view. During this same period, the Greek civilization was developing, and our present constellation names originated there. The Greeks borrowed from the Babylonians, but they went much further. The idea that natural events can be understood, and described numerically, originated with the teachings of Pythagoras. The realization that the moon shines by reflected sunlight, and thus the correct understanding of eclipses, dates from this period.

Plato's teachings, that understanding should come through reason rather than direct observation, dominated Western thinking for almost 2000 years. One of his students was Aristotle, who was the first to adopt physical laws to explain the Universe. With the development of geometry, it became possible for the Greeks to deduce such things as the relative sizes and distances of the moon and sun, and the size of the earth. Perhaps the greatest Greek astronomer was Hipparchus, who measured star positions, discovered precession of the stars, refined the methods of determining sizes through the use of trigonometry, and developed the magnitude system used today by astronomers. Greek astronomy ended about 150 A.D. with the work of Ptolemy, who compiled an encyclopedia of all astronomical knowledge.

During the Dark Ages in Europe, the Arabs kept Greek knowledge alive. One major contribution they made was the introduction of our present numbering system with its concept of zero. The Arabs developed algebra. Chinese astronomy also began to develop at the same time as Babylonian astronomy. Knowledge of timekeeping was well developed, although the Chinese attached little significance to planetary motions. By the first century A.D. they could predict eclipses and were aware of precession, but from then on, the Chinese culture was in contact with the Western world, making it difficult to separate the development of their astronomical knowledge from outside influence.

Because most of the writings of the Maya civilization in central America were destroyed by the Spanish, it is difficult to appreciate the achievements of this culture. It is known, however, that these people developed a very accurate calendar and measured planetary motion with some precision.

Names

Aristarchus
Aristotle
Eratosthenes
Hipparchus
Plato
Ptolemy
Pythagoras

Exercise: Sky Maps

Although there is no substitute for direct observation of the sky, a map or chart of the stars is a great help in orienting yourself and learning the constellations. The problems in constructing star maps are similar to the problems in mapping the earth onto a flat surface. When we observe the sky, we are looking at the inside of a dome where distances between objects are measured in angles.

You will need a map of the sky centered on the north pole and showing stars as far south as -40°, available in most stores that sell scientific supplies, or in planetariums and telescope shops. Such a chart is analogous to a polar projection, and consequently the edges suffer substantial distortion. On any particular night, you can only see some of these stars. Let us modify this map to show the sky on a particular date and time.

First let us find what stars pass through your zenith point. Note that declination is marked on the star map in 10°intervals, from the north pole to -30°. From your reading, you know how to find the declination of your zenith: it is your latitude on earth. Thus, over every 24 hour period, the stars at your zenith declination will pass through your zenith as they revolve around the north pole. This is fine, but you still need to know at what time of day or night a particular star will pass through your zenith. For this, you first need to know where the sun is on a particular date.

Note that the path of the sun is marked on the map, with the dates indicated also. At noon, the sun will by approximately on your meridian. (Do you want to review the definition of meridian?) To see what star is closest to your zenith at noon on any date, draw a line from the sun's position to the north pole and see where it intersects your zenith. For example, if you are at latitude 40°, Vega will pass through your zenith at noon on January 1. And if you want to know what the rest of the sky will look like at noon on January 1, remember that the observable sky is a hemisphere: you can see about 90°from the zenith in all directions. You might want to trace a circle on a piece of paper with a radius equal in length to 90°on the star map, cut out and discard the circle, and center the paper on the zenith point to see the sky on that date and time.

Now you can show what the sky looks like at noon on any day of the year, but what you really want is to see the sky at night. Your star finder can be set to any hour of the day or night if you remember that the stars appear to rotate every 24 hours, and that the stars at your zenith declination will always pass through your zenith, which is the center for the hole you cut out of the piece of paper. Mark the four cardinal points on your paper overlay. North is closest to the pole along the meridian, south is 180°away. To determine east and west, remember that they are reversed from a terrestrial map. Now suppose you want to see the sky at 6 P.M. on January 1. Rotate the sky map about the north pole through 90°, keeping the correct zenith point at the center of the cut-out circle. Remember that the stars should rise in the east and set in the west. You will find that at 6 P.M. the globular cluster M31 is at your zenith point. Any hour can be set in this manner by remembering that 1 hour equals a 15°rotation of the earth.

Now you can set your sky map for any date and time. Try the following exercises:

a. What constellation is overhead about midnight, June 1?
b. What time does the constellation Orion rise in October?
c. In what constellation did the sun appear on your birthday? This is known by astrologers as your sun sign. Is your answer the same as that given in an astrology column in a newspaper?
d. Draw the celestial equator on your star map. In what constellation is the sun on the vernal equinox? Was this true at the time of the Greeks?

What you have constructed is usually referred to as a planisphere. Commercial ones can be purchased at most planetariums.

Self Test

1. To an observer at the north pole
   a. the celestial equator passes directly overhead
   b. any star north of the equator will never set
   c. the sun never sets
   d. the days and nights are always 12 hours long
   e. every star in the entire sky is visible at some time during the year.

2. An observer on the earth's equator will see:
   a. stars move parallel to the horizon as the earth turns
   b. the entire celestial sphere in the course of a year
   c. the north celestial pole but not the south pole
   d. the sun pass through the zenith every day.

3. The Greeks of the first and second centuries were sure that the Earth was round because
   a. travelers reported a different appearance of the sky in more northern latitudes.
   b. the stars were noted to rise 4 minutes earlier each day than the day before
   c. the sun and moon were obviously round
   d. Copernicus had proved that the earth was a planet.

4. The effect of precession is to
   a. gradually change the angle or inclination of the earth's axis
   b. gradually change the shape and appearance of the constellations
   c. gradually change the constellations that define the zodiac
   d. gradually shift the intersection of the equator and ecliptic along the ecliptic.

5. According to Ptolemy, the planets moved on small circles that in turn moved along secondary circles called
   a. deferents
   b. epicycles
   c. syzygies
   d. parabolas.

6. The idea that the secrets of nature could be solved by reason alone was taught by
   a. Hipparchus
   b. Aristotle
   c. Ptolemy
   d. Plato.

7. The primary motivation for the Babylonians to study the sky was
   a. to determine the time of the spring floods
   b. to predict eclipses
   c. to make astrological predictions
   d. to establish time keeping techniques.

8. The first to assert that the earth was round was
   a. Pythagoras
   b. Aristotle
   c. Plato
   d. Ptolemy.

9. The cosmology of Plato and Aristotle included
   a. a flat earth at the center of the Universe
   b. a spherical earth at the center of the Universe
   c. a spherical sun at the center of the Universe
   d. a spherical earth at the center of the solar system.

10. The earliest recorded astronomical data were found in
    a. Stonehenge, England
    b. China
    c. Iraq
    d. Mexico.

ANSWERS TO SELECTED REVIEW QUESTIONS

1. Since $12 \times 30 = 360$, and since there are actually about 365 days in one year, the Babylonian calendar would have "fallen behind" by 5 days each year, so that in 6 years the calendar would be 30 days behind and an extra month would have to be added. Thus, the Babylonians had to add an extra month to their calendar every 6 years.

3. If the angle to the sun at Alexandria were 30° on the day that the sun were overhead at Syene, this would mean that Alexandria and Syene were $\frac{30}{360} = \frac{1}{12}$ of the earth's circumference apart. Then the circumference of the earth would be $12 \times 3300 = 39,600$ km. The diameter of the earth is just the circumference divided by π, so the diameter would be $\frac{39,600}{\pi} = 12,605$ km.

5. From simple geometry (and trigonometry) the formula for relating distance and parallax can be derived. If r is the distance in parsecs and p is the parallax in arcseconds, then

$$r = \frac{1}{p}.$$

If Hipparchus could only measure angles of 2 arcminutes or more, then p would have had to be at least 120 arcseconds, so

$$r = \frac{1}{120} = 0.008 \text{ parsecs.}$$

In order for Hipparchus to have been able to measure a parallax, the star would have had to be about 150 times closer than the nearest star, α Centauri!

Answers to Self Test

1-b, 2-b, 3-a, 4-d, 5-a, 6-d, 7-c, 8-a, 9-b, 10-c

Additional Readings

Aveni, A. 1979, Venus and the Maya, *American Scientist*, **67**:274.

Chamberlain, V. D. 1976, Prehistoric American astronomy, *Astronomy*, **4**(7):10.

Chamberlain, V. D. 1981, The Skidi Pawnee chart of the heavens, *Sky and Telescope*, **62**(1):23.

Daniel, G. 1980, Megalithic monuments, *Scientific American*, **243**(1):78.

Heath, T. L. 1969, *Greek Astronomy*. New York: AMS.

Hicks, R. D. 1976, Astronomy in the ancient Americas, *Sky and Telescope*, **51**(6):372.

Krupp, E. C. 1978, *In Search of Ancient Astronomies*. Garden City, N.J.: Doubleday.

Kunitsch, P. 1983, How we got our "Arabic" star names, *Sky and Telescope*, **65**(1):20.

Neugebauer, O. 1969, *The Exact Sciences in Antiquity*. New York: Dover.

Rodriguez, L. 1975, Ancient astronomy in Mexico and Central America, *Mercury*, **4**(1):24.

Stephenson, F. R. 1982, Historical eclipses, *Scientific American*, **247**(4):170.

Stephenson, F. R. and Clark, D. H. 1977, Ancient astronomical records from the Orient, *Sky and Telescope*, **53**(2):84.

Williamson, R. 1978, Native Americans were continent's first astronomers, *Smithsonian*, **9**(7):78.

Chapter 4

The Renaissance

With the re-awakening of an intellectual spirit in the fifteenth century, European astronomy began to develop. Progress can be identified with five individuals: Copernicus, Tycho Brahe, Kepler, Galileo, and Newton.

Nicholas Copernicus, encouraged by the climate of inquiry of his time, developed doubts about the geocentric view of the solar system. Since his heliocentric model required perfect circular motion, it was no more accurate in predicting planetary positions than the Ptolemaic system. Although Copernicus publicly disavowed that his model could be treated as anything but a mathematical tool for predicting planetary motion, it had profound impact on later scholars. Without better observational data, however, there was no test of Copernicus' theory, but fortunately for astronomy the observations were soon provided through the work of Tycho Brahe, a Dane. Although Tycho did not subscribe to the heliocentric theory, he was the first to make repeated systematic observations of planetary positions with a degree of accuracy far surpassing anything before then.

The analysis of Tycho's data was carried out by his assistant, Johannes Kepler. He began work with Tycho's data on Mars, and eventually determined that the orbit was an ellipse with the sun at one focus. This he later generalized as his first law of planetary motion. The second law he demonstrated was that a line connecting the sun and any planet sweeps out equal areas in equal time intervals. His third law stated that the square of the period of a planet is proportional to the cube of the semimajor axis of its orbit. As a result of his work, Kepler was able to construct tables for the prediction of planetary motion which were almost 100 times more accurate than previous tables.

Galileo Galilei, a contemporary of Kepler's, was an important experimental physicist whose astronomical discoveries proved the validity of the Copernican system. He was the first to turn the recently invented telescope to the sky, and he rapidly discovered lunar craters, the satellites of Jupiter, the phases of Venus and the existence of sunspots.

Words and Phrases

astronomical unit
ellipse
focus, foci (pl.)
harmonic law
mass
semimajor axis

Exercise: The Phases of Venus

Galileo's telescopic discovery that Venus shows phases like the moon had profound significance for the Copernican view. The observation could not be made without a telescope, but the Ptolemaic theory would have predicted something quite different. Remember that in the Ptolemaic system, Venus orbited the earth interior to the orbit of the sun. Describe what Ptolemy would have expected Venus to look like through a telescope over the course of its synodic period.

Self Test

Matching Exercise

1. Ecliptic
2. heliocentric
3. solstice
4. equinox
5. zodiac
6. epicycle
7. ellipse
8. meridian
9. semimajor axis
10. zenith
11. synodic period
12. eclipse

a) great circle passing through the north celestial pole and the zenith

b) circle along which planets and sun move in the geocentric cosmology

c) point directly overhead

d) device in geocentric cosmology to explain retrograde motion

e) event in which the moon blocks the sunlight as seen from the earth

f) point where sun crosses the celestial equator going north or south

g) centered on the sun

h) the axis drawn through the foci of an ellipse

i) orbital period of a body with respect to a distant fixed body

j) constellations along the ecliptic

k) orbital period of a body with respect to the sun

l) point where the sun reaches the maximum distance north or south of the celestial equator

m) geometric figure related to a circle, but having two foci rather than one

n) apparent path of the sun in the sky

ANSWERS TO SELECTED REVIEW QUESTIONS

1. The total circumference of the circle is given by

$$C = 2\pi R = 63 \text{ ft},$$

so the distance along the quarter circle is $\frac{1}{4} \times 63$ feet, or 15 feet, 9 inches (189 inches). If there were 90°, then each degree mark must have been $\frac{1}{90} \times 189$ inches, or 2.1 inches apart. Since there are 60 minutes in one degree, the minute marks must have been $\frac{1}{60} \times 2.1$ inches, or .035 inches apart.

2. The average of these values is 27° 14′, to the precision of the measurements. Thus the average value is equal to the true value, within the accuracy of the measurements. If a single measurement is made, random errors in the measurement process will affect how accurate the measurement is. But if several measurements are made and then averaged together, the effects of random measurement errors can be minimized, and the resulting value is much more accurate than any single measurement. This is one reason why Tycho Brahe's measurements were so much more accurate than those of earlier astronomers.

5. Since $P^2 \propto a^3$, the ratio of $\frac{P^2}{a^3}$ must always be the same. In units of A.U. and years, that ratio must be 1, because it is 1 for the Earth. We can then always solve the equation

$$\frac{P^2}{a^3} = 1, \ \text{or} \ P^2 = a^3.$$

Therefore, for a planet orbiting at 3 A.U.,

$$P = (3^3)^{\frac{1}{2}} = 27^{\frac{1}{2}} = 5.2 \text{ years}.$$

For a planet whose period is 11.18 years,

$$a = ((11.18)^2)^{\frac{1}{3}} = (125)^{\frac{1}{3}} = 5 \text{ A. U.}$$

Answers To Self Test

1–n, 2–g, 3–l, 4–f, 5–j, 6–d, 7–m, 8–a, 9–h, 10–c, 11–k, 12–e

Additional Readings

Numerous helpful articles about the people discussed in this chapter and their achievements can be found in the books listed in the text, and in such sources as the Journal for the History of Astronomy and the Dictionary of Scientific Biography. Here are some articles from popular periodicals:

Byard, M. M. 1977, Poetic responses to the Copernican revolution, *Scientific American*, **236**(6):120.

Drake, S. 1973, Galileo's discovery of the law of free–fall, *Scientific American*, **228**(5):84.

Drake, S. 1975, The role of music in Galileo's experiments, *Scientific American*, **232**(6):98.

Drake, S. 1976, Galileo and the first astronomical computing device, *Scientific American*, **234**(4):104.

Drake, S. 1980, Newton's apple and Galileo's Dialogue, *Scientific American*, **243**(2):150.

Drake, S. and Kowal, C. T. 1980, Galileo's sighting of Neptune, *Scientific American*, **243**(6):74.

Drake, S. and MacLachlan, J. 1975, Galileo's discovery of the parabolic trajectory, *Scientific American*, **232**(3):102.

Gingerich, O. 1973, Copernicus and Tycho, *Scientific American*, **229**(6):86.

Gingerich, O. 1977, Tycho Brahe and the great comet of 1577, *Sky and Telescope*, **54**(6):452.

Gingerich, O. 1982, The Galileo affair, *Scientific American*, **247**(2):132.

North, J. D. 1974, The astrolabe, *Scientific American*, **230**(1):96.

Wilson, C. 1972, How did Kepler discover his first two laws?, *Scientific American*, **226**(3):92.

Chapter 5

Isaac Newton and the Laws of Motion

Sir Isaac Newton certainly ranks as one of the foremost scientists of all times, and his discoveries in astronomy alone form the basis for much of our understanding of the mechanics of the heavens. Expanding on work by Galileo, Newton developed the concept of mass (and the related concepts of weight and inertia), and force, acceleration and velocity. His three laws of motion summarize these ideas:

1) a body at rest or in uniform motion tends to stay that way unless a force acts on it;
2) when a force is applied to a body, it is accelerated, and the acceleration is equal to the force divided by the mass of the object;
3) forces occur in equal and opposite pairs.

In order to explain the force that accelerates the planets around the sun, Newton used Kepler's third law. He also showed that a similar force described both the acceleration of the moon around the earth and the measured acceleration of a falling body, and formulated his law of universal gravitation: every pair of bodies is attracted by a force that is proportional to the masses and inversely proportional to the square of the distance between them. Newton's laws of motion and gravitation led him to modify Kepler's laws slightly. In particular, he showed that when two bodies orbit a common center of mass, the period squared times the sum of the masses of the bodies equals the cube of the semi-major axis, whereas Kepler had been unaware of the dependence on the sum of the masses. This law then became a powerful tool for determining masses of distant objects.

Newton's concept of energy led to the idea of escape velocity, which is applied today to many problems, from rocket launches to the escape of planetary atmospheres. Newton was also able to explain the tidal force, or differential gravitational force. For extended bodies which are relatively close to one another, the gravitational force changes relatively rapidly with distance, so the force changes along the diameter of the body. This differential force can slow the body's rotation and even lead to its break-up. In the case of the earth, the differential force causes the oceans to flow and creates high tides on opposite sides.

Words and Phrases

acceleration
angular momentum
aphelion
ellipse
energy, potential and kinetic
escape velocity
focus, foci (pl.)
force
gram
harmonic law
inertia
law of gravitation
luminosity
mass
perihelion
tidal force
weight

Exercise: Learning to Like Equations

Many students, when confronted with an equation, have a tendency to panic. Even if they know which equation applies to the problem at hand, and their algebra is up to the necessary manipulations, they still feel unable to tackle the problem. Let's discuss some of the reasons for this. Many of the physical quantities which we have been dealing with are unfamiliar and difficult to visualize, like force and momentum. Then there is the question of what units to use. Should you use miles or kilometers? Will it make a difference if you use grams or kilograms?

There are several important ideas about equations which will make working with them easier, and allow you to check your work and become more confident. First, an equation such as the law of gravity must be true no matter what units are used to measure the physical quantities. Therefore an equation may be solved using different sets of units. However, it is important to realize that the units must balance. An equation is like a seesaw: the two sides are equal by definition. But unlike the seesaw in the park where you can balance your kid brother and a sack of potatoes, an equation demands the same quantities on both sides. This concept can be very useful in solving problems and in checking your solution. If you find that you have set kid brothers equal to potatoes, something is wrong.

As an example, consider the relation $F = m \times a$. The units on the right hand side could be mass m, expressed in grams, and acceleration a, expressed in cm/sec^2. In that case the force has the units of $gm \times (cm/sec^2)$. This unit of force has a name; it is called a dyne. In working any problem in which the force is to be expressed in dynes, the unit of mass must be grams and the unit of acceleration must be cm/sec^2. Now suppose that in the same equation the units of mass are kilograms and the units of acceleration are $kilometers/sec^2$. What are the units of force?

Occasionally an equation contains a constant, sometimes referred to as a constant of proportionality. An example is the constant G in the law of gravitation. Since all of the other quantities in the law have physical dimensions, the constant must also have physical dimensions, and its value will depend on the set of units used to measure the other quantities in the equation. This is where a constant like G differs from a dimensionless constant like π, the ratio of the circumference to the diameter of a circle. If you are not sure if a constant has dimensions or not, try balancing the units in the equation. If they balance without any help from the constant, then the constant is dimensionless. For example, consider the formula for the area of a circle, $A = \pi r^2$. If r is measured in meters and area is measured in square meters, then π is dimensionless.

If you pay attention to balancing the units in an equation, it will be clear which ones should be used. Try the following problems for practice.

1. In the law of gravitation, there is a constant usually abbreviated as G. Determine the units that this constant has if the force is in units of dynes, the mass in grams and the distance in centimeters. (Note that $F = Gm_1m_2/r^2$, so $G = Fr^2/m_1m_2$.

2. Suppose that you are faced with calculating the gravitational force in dynes between the earth and the moon. You are given the masses in grams and the earth–moon distance in miles. What must you do to the units before solving the equation?
3. If the escape velocity of a planet is given by $v_e = 2GM/R$, what are the units of v_e? Assume M is the planet's mass in grams, R is its radius in cm, and that you have found the units for G in question 1.

Self Test

1. The earth's gravitational force extends as far as
   a. the moon
   b. the sun
   c. the edge of the solar system
   (d.) the edge of the universe.

2. Kepler's discovery that planetary orbits are really ellipses, rather than circles
   a. is entirely consistent with Newton's theory of gravitation
   b. eliminated the need for epicycles and deferents, making calculations of planetary positions easier and more accurate
   (c.) both of the above
   d. none of the above.

3. If we double the distance of an artificial satellite that has been in a synchronous orbit around the earth, its new period will be
   a. 1 day
   b. 1.6 days
   c. 2 days
   (d.) 2.8 days.

4. An astronaut in a spaceship orbiting the earth at a height of 500 km
   a. is too far from the earth to feel the effect of gravity
   b. is falling towards the floor of his spacecraft
   c. weighs only one–fifth his ground weight
   (d.) none of the above.

5. The true diameter of a planet may be determined when two of the following are known (choose two):
   a. the mass of the planet
   b. the radius of the planet's orbit in kilometers
   c. the planet's rotation period
   d. the angular size of the planet's observed disk.

6. The ability of a planet to maintain an atmosphere depends on
   a. its mass and rotation period
   b. its temperature and number of satellites
   c. its escape velocity and the velocity of its gas molecules
   d. its magnetic field.

7. If you went to the moon, which of your physical characteristics would be guaranteed to change?
   a. mass
   b. height
   c. weight
   d. density.

8. How many of the following are examples of acceleration?
   a. a car going 60 km/sec down a straight road
   b. a car turning a corner
   c. a wagon rolling down a hill
   d. the earth orbiting the sun.

9. If the law of gravitation stated that the force between two bodies was inversely proportional to the cube of the distance between them,
   a. the force would be weaker between the sun and earth than it is now
   b. the force would be stronger between the sun and earth than it is now
   c. the force would be the same at the distance of the earth, but weaker for planets farther from the sun
   d. there is no way to tell without an experiment.

10. The escape velocity of a planet depends in its
   a. mass and temperature
   b. density and temperature
   c. mass and radius
   d. none of the above.

ANSWERS TO SELECTED REVIEW QUESTIONS

Problems

1. Suppose that at sea level you weigh 130 lb. 1 kg = 2.2 lb, so your mass would 130/2.2 = 59 kg.

3. The gravitational force between a planet and the Sun is given by

$$F_{\text{grav}} = \frac{Gm_{\odot}m_{\text{p}}}{r^2}.$$

Since we are only interested in the ratios, we can simplify the equation to

$$Ratio = \frac{m_{\text{p}}}{r_{\text{p}}^2}$$

where m_p and r_p are in the units $M_{\oplus}$ and AU; the ratios for Saturn and Jupiter are:

Planet	*Mass*	*Distance*	*Ratio*
Earth	1	1	1
Jupiter	300	5	12
Saturn	100	10	1

5. We can again use a ratio approach. From Appendix 7, we see that Titan has a radius of 0.4 Earth radii, and a mass of 0.02 Earth masses. From this the gravitational force on Titan must be $0.02/(0.4)^2 = 0.14$ times as large, so if you weighed 130 pounds on the surface of the Earth, you would weigh $130 \times 0.14 = 18$ pounds on the surface of Titan.

7. 1 watt is defined as 1 joule per second, and 1 joule is equal to 10^7 ergs, so 1 watt = 10^7 ergs per second. Thus a 100 watt light bulb emits 100×10^7 ergs per second, or 10^9 ergs per second.

Answers To Self Test

1–d, 2–c, 3–d, 4–d, 5–b d, 6–c, 7–c, 8–b c d, 9–a, 10–c

Additional Readings

The laws of physics and gravitation are not often featured in articles in popular magazines and periodicals, but there are many elementary textbooks available in any science library in which these principles are clearly explained, and examples are given.

There do exist popular articles on Newton:

Byard, M. M. 1977, Poetic responses to the Copernican revolution, *Scientific American*, **236**(6):120.

Cohen, I. B. 1981, Newton's discovery of gravity, *Scientific American*, **244**(3):166.

Drake, S. 1973, Galileo's discovery of the law of free–fall, *Scientific American*, **228**(5):84.

Drake, S. 1980, Newton's apple and Galileo's Dialogue, *Scientific American*, **243**(2):150.

Chapter 6
The Nature of Light

What we refer to as light is actually only one slice from the middle of a long loaf. Since all our information about stars comes from the loaf, it requires careful understanding. In some experiments, light acts much like a wave, with red light exhibiting longer wavelengths than blue light. Outside the range of our vision, increasing wavelength includes infrared, microwave and radio waves and decreasing wavelength covers ultraviolet and gamma rays. The full spread of possible wavelengths is called the electromagnetic spectrum.

Experiments show that electromagnetic waves will bend around sharp corners and interfere with each other when their crests are out of phase, just as other types of waves do. In other experiments, light acts as though it consists of particles. Light can travel in a vacuum, and it carries energy in discrete packets called photons. The wavelength and the energy of a photon are inversely proportional to one another, and the photon energy can be absorbed by matter. When we consider the overall light emitted by an object, including those parts of the spectrum to which our eyes are insensitive, we are dealing with a continuous spectrum. The amount of energy at each wavelength is determined by the temperature of the object. Wien's law relates the temperature and wavelength of maximum emission for a glowing object, stating that the two quantities are inversely proportional to each other. The Stefan-Boltzmann law says that the energy emitted per square centimeter of surface area is proportional to the fourth power of the temperature; this may be combined with the surface area of a sphere to describe the total energy emitted over the surface of a star as being proportional to the square of the radius and the fourth power of the temperature.

If we use a prism to sort visible sunlight into its different wavelengths, we find that at some particular wavelengths there is no light. We refer to these as absorption lines. In another case, we view a hot, rarefied gas and observe only a few bright emission lines rather than a continuous spectrum. These and other spectroscopic observations can be explained by considering light as a particle and exploring its behavior with respect to an atom. An atom can be thought of as a solar system in miniature with electrons orbiting the nucleus. Each possible electron orbit corresponds to a specific amount of energy. An electron can absorb only a photon with the exact energy required for the electron to make a transition from one orbit to another. Absorption of photons results in absorption lines, and re-emission of these photons produces emission lines. The conditions under which absorption and emission lines are formed are summarized in Kirchhoff's rules.

The pattern of emission or absorption lines is unique for each chemical element, and this allows us to determine the composition of stars. Since the pattern changes as the atoms are excited and ionized, we can also study the density and temperature of a star's atmosphere. Another important source of information in stellar spectra is the Doppler shift. When a star is moving towards or away from us, the entire spectrum will be shifted either to the blue or the red by an amount proportional to the speed of the star along our line of sight. The shift of the continuous spectrum is very difficult to measure, but if there are absorption or emission lines in the spectrum, the velocity can be directly measured.

Words and Phrases

atom
Angstrom
absorption line
blue-shifted spectrum
continuum
diffraction
Doppler shift
electromagnetic spectrum
electron
emission line
energy levels
excitation
frequency (of radiation)
gamma rays
infrared radiation
ion
ionization
Kirchhoff's rules
microwaves
nucleus
photon
radio waves
red-shifted spectrum
spectroscopy
spectrum
Stefan-Boltzmann law
thermal radiation
thermal radiation
ultraviolet radiation
wavelength
Wien's Law
X-rays

Exercise: The Temperature of the Sun

From the principles in Chapter 5 and a very simple experiment that requires only a thermometer, a pan of water, and a warm day, it is possible to determine the surface temperature of the sun with surprising accuracy. The total luminosity of the sun is just the energy emitted at the sun's surface per unit area per unit time, multiplied by the total area of the sun. The energy is related to the sun's temperature by the Stefan-Boltzmann Law. We can express the luminosity, therefore as:

$$L_s = \sigma T_s^4 (4\pi r_s^2),$$

where r_s is the radius of the sun. At the earth's distance from the sun, this luminosity has spread to fill a much larger volume, but no energy has been lost. We can now express the luminosity as:

$$L_e = \sigma T_e^4 (4\pi r_e^2),$$

where T_e is now the temperature at the distance r_e of the earth from the sun. Since no energy has been lost, we can set the equations for L_s and L_e equal, and we find that a few quantities cancel out:

$$T_s^4 r_s^2 = T_e^4 r_e^2.$$

Assuming that we know the radius of the sun and the distance to the sun, all we need to do is measure the temperature T_e, and we can solve for T_s, the temperature of the sun.

Here is where the pan of water comes in. If you take it out in the sun on a warm, quiet day, and put a few drops of black ink in the water to help it absorb heat, the water temperature will reach equilibrium. There are some energy losses which prevent our using the measured temperature directly, however. In the first place, the earth's atmosphere screens out some of the incoming radiation, and in the second place, our pan of water is not able to absorb all the radiation that hits it. If we assume that the atmosphere absorbs 65% and the pan of water absorbs 75%, then our relation should be expressed as:

$$T_s^4 r_s^2 = \frac{T_e^4 r_e^2}{(.65) \times (.75)}.$$

Find the temperature of the sun and compare it with the value in the text. Remember that the text gives it in units of K (that is, on the absolute temperature scale).

Self Test

1. Which of the following is not a type of electromagnetic radiation?
   a. visible light
   b. radio waves
   c. x-rays
   d. sound waves
   e. infrared energy.

2. To determine what kinds of atoms the outer regions of a star are made of, astronomers would measure the star's
   a. continuous spectrum
   b. distance
   c. spectral absorption lines
   d. luminosity
   e. radius.

3. Our sun emits its most intense radiation in what region of the electromagnetic spectrum?
   a. radio
   b. infrared
   c. visible
   d. ultraviolet
   e. x-ray.

4. The Doppler effect
   a. is a measure of a star's space velocity
   b. is a shift of the star's spectrum which depends on its velocity in the line of sight
   c. is a shift in the star's apparent position with respect to background stars
   d. is a shift in the star's spectrum which depends on its temperature.

5. The brightness of an object
   a. is inversely proportional to the distance between object and observer
   b. is inversely proportional to the square of the distance between object and observer
   c. is directly proportional to the square of the distance between object and observer
   d. none of the above.

6. In light of the same color, each photon has the same
   a. wavelength
   b. energy
   c. frequency
   d. all of the above.

7. An atom in the ground state
   a. has lost its electrons
   b. can emit a photon
   c. is in its state of greatest energy
   d. can absorb a photon
   e. none of the above.

8. Wien's Law states
   a. the cooler an object, the bluer the radiation it emits
   b. the hotter an object, the bluer the radiation it emits
   c. the hotter an object, the redder the radiation it emits
   d. force is equal to mass times acceleration.

9. The spectrum of an ionized atom
   a. lacks some of the lines seen in the neutral atom
   b. is different from the spectrum of the neutral atom
   c. is the same as the neutral atom
   d. resembles the spectrum of the next lightest element.

10. Spectral lines are produced by transitions between energy levels in
    a. atoms
    b. ions
    c. molecules
    d. all of the above.

ANSWERS TO SELECTED REVIEW QUESTIONS

Problems

1. The FM band has frequencies of megahertz. For your favorite, take your pick; that of one of the authors (KSB) is KBCO-FM at 97.3 MHz which has a wavelength given by $\lambda = c/\nu = 308.3$cm.

3. A wavelength of 4500Å is equivalent to a wavelength of 4.5×10^{-5} cm. A photon with this wavelength would have an energy of $E = hc/\lambda$, where $h = 6.62 \times 10^{-27}$ erg sec (Planck's constant) and c is the speed of light. Thus its energy would be

$$E = \frac{(6.62 \times 10^{-27}) \times (2.9979 \times 10^{10})}{4.5 \times 10^{-5}} = 4.4 \times 10^{-12} \text{ ergs.}$$

5. The energy per second per unit surface area is given by $E = \sigma T^4$. If the temperature of Mars is 5 times that of Pluto, then Mars will emit $5^4 = 625$ times more energy per unit surface area per second than Pluto.

7. Intensity of light falls off as one over the distance squared. If Pluto is 25 times as far from the Sun as Mars, then the amount of light reaching Pluto will be diminished by $1/(25)^2 = 1/625$. Thus, the amount of light reaching Mars from the Sun is 625 times more than reaches Pluto.

9. Using the technique in the previous problem, we find that the wavelength is 9.11×10^6 cm = 911 Å. From the equation for kinetic energy, we know that $E = \frac{1}{2}mv^2$. The mass of an electron is $m_e = 9.109 \times 10^{-28}$ grams. Solving for the velocity gives

$$v = \sqrt{\frac{2E}{m_e}} = 2.2 \times 10^8 \text{ cm/sec.}$$

Answers To Self Test

1–d, 2–c, 3–c, 4–b, 5–b, 6–d, 7–d, 8–b, 9–b, 10–d

Additional Readings

As in the case of the laws of motion and gravitation, there are few popular articles to be found on light and its properties. The best resources for further clarification of the concepts presented in this chapter are elementary textbooks available in any science library.

Chapter 7

Telescopes on Earth and in Space

From the time that Galileo first turned his telescope to the sky until about 1950, all observational astronomy was done through one narrow window of the electromagnetic spectrum: visible light. In the last few decades new technology has enabled us to study almost the entire spectrum from X-rays to radio waves. Different techniques are needed to collect and analyze the light from different spectral regions, but some principles are the same.

A telescope is a light bucket, a device that can collect far more photons than the human eye. The instrument attached to it can sort these photons out if the astronomer desires and then store them by means of film or some type of electronic detector. The most common telescope design, from the UV through the optical, infrared and radio region, is the reflector. The light is reflected from a concave mirror, either spherical or parabolic, to a point in front of the mirror. In some cases this is the focus of the mirror and the recorder is placed there. In other designs the light is reflected off another mirror either back through a hole in the primary mirror or off to the side of the light path.

Another somewhat less common design for professional telescopes is the refractor, which employs a lens as light-collecting element. These are technically difficult to build beyond about 1 meter in diameter, so they are most common in small telescopes. Their main advantage is in photographing the sky on a wide scale, which is necessary for positional measurements.

In addition to directly recording the image produced by the telescope, a spectrograph is often used to examine the spectral distribution of the star light. Other times the light is recorded after it passes through different filters to measure the intensity in different spectral regions. Such a device is called a photometer. In all cases the image or spectrum is stored on film or electronically for later study. Since most exposures with a telescope last from several minutes to hours, it is necessary for the telescope to track the object and compensate for the earth's rotation. Sites for earth-based telescopes are chosen on the basis of their clear, steady air and absence of light pollution. Increasingly, telescopes are being put in orbit where they are above all of the earth's atmosphere and therefore can observe in any wavelength region.

Words and Phrases

cassegrain focus
chromatic aberration
coude focus
eyepiece
focal length
focus (of lens)
interferometry
Newtonian Focus
photometer
prime focus
reflecting telescope (reflector)
refracting telescope (refractor)
resolution
spectrograph

Exercise

In understanding lenses and telescopes, it is helpful to think about one "telescope" that most people have used: the camera. A camera generally is designed along the lines of a refracting telescope, so that light passes through a lens and is focused ultimately on film to record the picture. The following exercises can be done with most cameras, in case you want to verify your answers.

Imagine that you have a camera in which the lens can be left open to take long exposures. On a clear night you set it up pointing at the north star and expose the film for three hours. What will your developed picture look like? Now suppose you point the camera at the celestial equator and take another three-hour exposure. How will this picture differ from the last one? Finally, suppose you attach your camera to the tube of a large telescope with a clock drive. If the camera is again pointing toward the celestial equator and you take a three-hour exposure, what will the picture look like?

Suppose your camera lens has the misfortune to be splattered with a drop of black paint which covers about one quarter of the lens. Describe how the same scene photographed before and after this accident would look (if you want to try this experiment, use a cardboard wedge to cut off part of the image without touching the lens!)

Self Test

1. Why can radio telescopes work during the daytime?
    a. radio waves are invisible
    b. the sun emits very few radio waves in comparison to visible light
    c. radio astronomers don't use photographic techniques
    d. radio waves are not scattered like visible light.
2. The main function of a telescope is generally to
    a. magnify star images
    b. separate light into its spectrum
    c. help us see radiation that cannot get through clouds
    d. collect more radiation from faint objects.
3. When you look at a star through a large telescope, you see
    a. a uniform disk
    b. a point of light
    c. a disk with surface features
    d. a spectrum.
4. Why do we place telescopes on satellites in earth orbit to measure celestial x-rays?
    a. to escape interference from artificial sources of x-rays
    b. because celestial x-rays cannot reach the ground
    c. celestial x-ray observations are too dangerous from the ground
    d. no one has observed x-rays from space.

5. The operation of the human eye is analogous to that of
   a. a reflecting telescope
   b. a coude focus
   c. a refracting telescope
   d. a radio telescope.

6. Stars twinkle because
   a. they are spheres of hot gas
   b. atoms are continually being excited in their atmosphere
   c. the earth's atmosphere distorts their light
   d. clouds of interstellar gas are continuously in motion in front of them.

7. The fact that visible light penetrates through the earth's atmosphere, but X-rays do not, indicates that
   a. X-ray photons are less energetic than visible light
   b. the opacity of air is much greater at X-ray wavelengths
   c. visible light, like neutrinos, has virtually no interaction with gas molecules
   d. the human eye is sensitive to visible light but not to X-rays.

8. Major observatories are usually located at high altitude because
   a. this gets the telescopes above some of the atmosphere's obscuring effects
   b. it reduces gravitational distortions in the mirror shapes
   c. it gets the telescopes closer to the target objects
   d. it allows X-rays to be observed from the ground.

9. If you double the diameter of a telescope, you:
   a. can resolve objects at one half their previous angular separation
   b. double the amount of light collected
   c. increase the magnification by a factor of four
   d. increase the amount of light collected by a factor of four.

Problems

1. The light gathering power is proportional to the area of the telescope aperture (*i.e.* the lens or mirror). The area is given by $A = \pi r^2$. To compare the light gathering capabilities of two telescopes, we can use the ratio:

$$\frac{A_1}{A_2} = \frac{r_1^2}{r_2^2}.$$

For the 6-m (236 inch) telescope and the 40 inch telescope, this gives

$$\text{Ratio} = \frac{(118)^2}{(20)^2} = \frac{13920}{400} = 34.8,$$

so the 6-m telescope has 34.8 times the light gathering power of the 40 inch telescope.

3. The fraction of light that is blocked by the cage is

$$\frac{(20)^2}{(100)^2} = \frac{1}{25} = 0.04 = 4\%.$$

5. We can again use a ratio method so that the constants cancel out. Then,

$$\frac{R\ (100\ \text{cm})}{R\ (10\ \text{cm})} = \frac{(\lambda_{100}/D_{100})}{(\lambda_{10}/D_{10})}.$$

Since the telescopes are observing at the same wavelength, the λ values cancel, and we find that the 1 meter telescope, with 10 times the diameter, can resolve objects at 0.1 arcsecond, which is 10 times better than the 10 cm telescope.

7. The equation for relating frequency and wavelength is

$$\nu = \frac{c}{\lambda}.$$

A photon with a wavelength of 21 cm would therefore have a frequency of 1.4×10^9 Hz, or 1400 MHz.

9. The diameter of the Space Telescope is 2.4 m. In order to make out surface markings on Pluto, which has an angular diameter of 0.1 arcseconds, the resolution must be better than 0.1 arcseconds. For a telescope the diameter of the Space Telescope observing at 5000 Å, the resolution would be 0.05 arcseconds, so the Space Telescope will be able to resolve surface markings on Pluto.

Answers To Self Test

1–b d, 2–d, 3–b, 4–b, 5–c, 6–c, 7–b, 8–a, 9–a d

Additional Readings

Bahcall, J. N. and Spitzer, L. 1982, The Space Telescope, *Scientific American*, **247**(1):40.

Carleton, N. P. and Hoffman, W. M. 1978, The Multiple-Mirror Telescope, *Physics Today*, **31**(9):30.

Cornell, J. 1979, On a mountaintop in Arizona, things are starting to look up, *Smithsonian*, **10**(2):42.

Fichtel, C., Greisen, K., and Kniffen, D. 1975, High-energy gamma-ray astronomy, *Physics Today*, **28**(9):42.

Giacconi, R. 1980, The Einstein X-ray Observatory, *Scientific American*, **242**(2):70.

Harnden, F. R., Jr. 1980, X-ray telescope in space reveals unseen universe, *Smithsonian*, **11**(6):110.

Harrington, S. 1982, Selecting your first telescope, *Mercury*, **11**(4):106.

Ioannisiani, B. K. 1977, The Soviet 6-meter altazimuth reflector, *Sky and Telescope*, **54**(5):356.

Isaacman, R. 1979, Using earth as a radio telescope, *Astronomy*, **7**(10):6.

Kellerman, K. I. 1972, Intercontinental radio astronomy, *Scientific American*, **226**(2):72.

Kristian, J. and Blouke, M. 1982, Charge-coupled devices in astronomy, *Scientific American*, **247**(4):56.

Labeyrie, A. 1982, Stellar interferometry: a widening frontier, *Sky and Telescope*, **63**(4):334.

Leventhal, M. and MacCallum, C. J. 1980, Gamma-ray line astronomy, *Scientific American*, **243**(1):62.

Maran, S. P. and Boggess, A. 1980, Ultraviolet astronomy enters the eighties, *Physics Today*, **33**(9):40.

Mims, S. S. 1980, Chasing rainbows: the early development of astronomical spectroscopy, *Griffith Observer*, **44**(8):2.

Overbye, D. 1979, The X-ray eyes of Einstein, *Sky and Telescope*, **37**(6):527.

Readhead, A. C. S. 1982, Radio astronomy by very-long-baseline interferometry, *Scientific American*, **246**(6):52.

Spitzer, L. 1978, The Space Telescope, *American Scientist*, **66**:426.

Stencel, R., Blair, W., and Conat-Stencel, S. 1978, Astronomical spectroscopy, *Astronomy*, **6**(6):6.

Strom, S. E. 1983, Infrared and submillimeter astronomy from space, *Sky and Telescope*, **65**(4):312.

Thaddeus, P. 1982, Facilities for U. S. radioastronomy, *Physics Today*, **35**(11):36.

Underwood, J. H. 1978, X-ray optics, *American Scientist*, **66**:476.

Wampler, E. J. 1982, Ultraviolet, optical, and infrared astronomy, *Physics Today*, **35**(11):44.

Chapter 8

The Earth as a Planet

A study of the earth, as compared with the other planets, offers an amazing contrast. Because the earth is the only planet with liquid water, a nitrogen-oxygen atmosphere, and of course, life, it is important to ask what factors have lead to this situation. Apart from these major contrasts, there are also similarities between the earth and the other planets, and a close look at our home planet aids us in understanding others in the solar system.

The earth's atmosphere has four distinct temperature zones, created by solar heating, which is maximized at the surface and in the ozone layer. Motions of the atmosphere are created by a combination of convection due to this solar heating, and rotation of the earth. Indirect studies of the earth's interior suggest a dense core of molten iron, a less dense mantle and a crust of silicates. This density gradient indicates that the earth has undergone a process called differentiation, as have some other bodies in the solar system. Part of the core appears to undergo fluid motions, and these create the earth's magnetic field. This field protects the surface of the earth from cosmic rays and other high-energy particles capable of causing genetic mutations. At intervals of thousands of years, however, the field decays and reverses polarity and at those times cosmic rays can reach the surface.

Convection currents in the mantle are responsible for continental drift. According to the theory of plate tectonics, the crust is made up of large moving plates that continually collide with one another, giving rise to earthquakes, volcanos, and the uplifting of mountains. The surface of the earth is rock, recycled from one form to another through erosion processes: water, wind, glaciation and biologic activity. The time since a rock formed can be measured through radioactive dating techniques, and ranges from millions to billions of years. Based on the oldest rocks on earth and other evidence, the age of the earth is placed at about 4.5 billion years.

The atmosphere of the earth is largely the result of biological activity on the surface. The most volatile gases, such as hydrogen and helium, escaped from the earth early in its existence. Plant life, along with volcanic outgassing, has replaced these gases with nitrogen and oxygen. The atmosphere also contains traces of water vapor, ozone and carbon dioxide, all very important for the continuation of life.

Words and Phrases

aerosols
albedo
asthenosphere
aurora borealis
carbon dioxide
compressional waves
continental drift
convection
core
crust
differentiation
igneous rock
ionosphere
lithosphere
magnetic dynamo
magnetosphere
metamorphic
mesosphere
oxides
ozone
P-waves
paleomagnetism
planetesimals
S-waves
scattered light
sedimentary
silicates
stratosphere
subduction
thermosphere
transverse waves
troposphere
volatile gases

Exercise: The Earth's Atmosphere

The thin layer of protective atmosphere above us is something which we often take for granted. It helps to put it in perspective by making a scale model of the earth and its atmosphere.

The radius of the earth is nearly 7,000 km. The atmosphere extends upward about another 80 km, or about 1% of the earth's radius. Let us scale this so that the earth has a radius of 100 cm. On a large sheet of paper, draw part of a circle with a radius of 100 cm. (a string tied to a pencil will make this easier). Above this arc representing the surface of the earth, draw another arc with a radius of 101 cm. This represents the atmosphere of the earth. Mark the height of Mt. Everest: 9 km, or about 1 mm on your scale.

Now consider the following problem. When you look straight up, you are looking through one atmosphere. How many times more atmosphere do you look through when you look at a star 45° from the zenith? How many atmospheres in the case of the setting sun?

Self Test

1. The present atmosphere of the earth is different from the atmosphere of a few billion years ago in that
   a. there was no water vapor until the oceans started to evaporate
   b. there was more ozone then because nobody was using deodorant or hair spray
   c. oxygen has been added in the last few billion years by the photosynthetic action of green plants
   d. there was no nitrogen until biological processes began.

2. The earliest evidence for life on earth dates back
   a. 3 million years
   b. 100 million years
   c. 3 billion years
   d. 1 billion years

3. From the study of continental drift, we conclude that the American and the European-African system have been separated from each other for about
   a. 50 million years
   b. 150 million years
   c. 700 million years
   d. 150 billion years.

4. The earth's atmosphere is composed mostly of
   a. oxygen
   b. carbon dioxide
   c. nitrogen
   d. hydrogen.

5. Which of the following types of rocks require erosion for their formation?
   a. igneous
   b. sedimentary
   c. basaltic
   d. none of the above.

6. The magnetic field of the earth
   a. is a consequence of biologic activity
   b. protects the earth from cosmic rays
   c. is radioactive and dangerous for astronauts to fly through
   d. acts as a shield against ultraviolet sunlight.

7. The earth's magnetic axis reverses at intervals
   a. regularly every 590 thousand years
   b. irregularly in tens of thousands of years
   c. regularly every million years
   d. irregularly over hundreds of millions of years.

8. The density of the continents is about
   a. 1 gm/cm^3
   b. 3.5 gm/cm^3
   c. 5.5 gm/cm^3
   d. 15 gm/cm^3.

9. Most of the weather processes take place at altitudes
   a. below 10 km
   b. between 10 and 50 km
   c. between 50 and 80 km
   d. below 2 km.

Problems

1. Density is defined as mass divided by volume. The volume of a sphere is $\frac{4}{3}\pi R^3$. The radius of the Earth is 6378 km = 6.378×10^8 cm, so the volume of the Earth is

$$\frac{4}{3}\pi(6.378 \times 10^8)^3 = 1.09 \times 10^{27} \text{ cm}^3.$$

The mass of the Earth is 5.976×10^{27} gm, so the density of the Earth is

$$\rho = \frac{5.976 \times 10^{27}}{1.09 \times 10^{27}} = 5.5 \text{ gm/cm}^3,$$

which agrees with the density listed.

3. If the core makes up 0.16 of the volume, then the total density is given by

$$\rho = 0.16\rho_{\text{core}} + 0.84\rho_{\text{outer}}.$$

Since $\rho = 5.5$ g/cm^3 and $\rho_{\text{outer}} = 3.5$ gm/cm^3, we solve for ρ_{core} and find that $\rho_{\text{core}} = 15.9$ gm/cm^3.

5. The half-life is the time required for half of the material to decay away. Therefore, if half the element is gone, the rock sample must be one half-life (5 million years) old. If only one-fourth the element remains, the rock is two half-lives (10 million years) old. If 1/64 of the element remains, the rock is 6 half-lives (30 million years) old.

Answers to Self-Test

1–c, 2–c, 3–b, 4–c, 5–b, 6–b, 7–b, 8–b, 9–a

Additional Readings

Alexander, T. 1975, A revolution called plate tectonics has given us a whole new earth, *Smithsonian*, **5**(10):30.

Alexander, T. 1975, Plate tectonics has a lot to tell us about the present and future earth, *Smithsonian*, **5**(11):38.

Ben-Avraham, Z. 1981, The movement of continents, *American Scientist*, **69**:291.

Bolt, B. A. 1973, The fine structure of the earth's interior, *Scientific American*, **228**(3):24.

Carrigan, C. R. and Gubbens, D. 1979, The source of the earth's magnetic field, *Scientific American*, **240**(1):118.

Dewey, J. F. 1972, Plate tectonics, *Scientific American*, **226**(5):56.

Fawl, H. 1978, A history of geologic time, *American Scientist*, **66**:159.

Hallam, A. 1972, Continental drift and the fossil record, *Scientific American*, **227**(5):56.

Hallam, A. 1975, Alfred Wegener and the hypothesis of continental drift, *Scientific American*, **232**(2):88.

Hartmann, W. K. 1978, The early history of planet earth, *Astronomy*, **6**(8):6.

Head, J. W., Wood, C. A., and Mutch, T. A. 1977, Geologic evolution of the terrestrial planets, *American Scientist*, **65**:21.

McKenzie, D. P. and Richter, F. 1976, Convection currents in the earth's mantle, *Scientific American*, **235**(5):72.

Molnar, P. and Tapponier, P. 1977, The collision between India and Eurasia, *Scientific American*, **236**(4):30.

O'Nions, R. K., Hamilton, P. J., and Evensen, N. M. 1980, The chemical evolution of the earth's mantle, *Scientific American*, **242**(5):120.

Rubin, A. E. 1980, The magnetic earth, *Griffith Observer*, **44**(1):2.

Siever, R. 1974, The steady state of the earth's crust, atmosphere, and oceans, *Scientific American*, **230**(6):72.

Siever, R. 1975, The earth, *Scientific American*, **233**(3):82.

Wetherill, G. W. 1981, The formation of the earth from planetesimals, *Scientific American*, **244**(6):162.

Wyllie, P. J. 1975, The earth's mantle, *Scientific American*, **232**(3):50.

Chapter 9

The Moon

For centuries the moon has served as a test for the laws of physics and our understanding of the solar system. In earlier times it provided a nearby example of the law of gravity, and the study of its motion led to a branch of astronomy known as celestial mechanics. In recent years it has become the only body in the solar system from which surface samples have been returned to the earth for close study. The moon proved a ready target for the space race, as the U.S. and Russia vied to be the first to orbit the moon, photograph the far side, and land instrument packages of increasing complexity on the lunar surface. Finally in 1969 the U.S. landed the first men on the moon, and followed this with five more manned landings over the next three years.

Even through a small telescope the surface of the moon is an impressive sight. Craters, maria and mountain ranges can easily be seen, the more so because the moon has no atmosphere to blur these features nor erode their jagged edges. The craters range in diameter from hundreds of kilometers to microscopic pits, testifying to a history of bombardment from space.

With all we have learned about the moon since the beginning of the space program, a clearer picture of its origin is emerging. Although the final answer is not yet in, it now appears that the earth and moon formed at the same time and have remained together ever since. As the earth coalesced from rocky chunks orbiting the sun, a ring of similar material may have formed about it. This ring quickly formed one body through collisions and capture, creating the moon, but leaving it much closer to the earth than it is at present. As the earth and the moon cooled, differentiation took place, but to a lesser degree in the moon than in the larger earth. Volcanic activity and cratering shaped the lunar surface during its early history. Collisions with fairly massive bodies triggered the massive lava flows that created the maria. This took place after much of the debris floating through the early solar system had been cleared out, and as a consequence the maria have not been as heavily cratered by impacts as the older portions of the lunar surface.

The moon has no magnetic field, and seismic experiments suggest a non-molten core, which is in keeping with its smaller mass compared with the earth.

Words and Phrases

craters (impact)
ejecta
fission
grazing occultation
maria
mascons
rays
regolith
rilles

Exercise: The Height of Lunar Mountains

The height of mountains on the moon can be computed from the shadows that they cast. All that is needed are some simple measurements and a little arithmetic. Find a picture of the moon which shows the shadows cast by craters on the moon, as well as the terminator line, the boundary between the daylit and dark sides of the moon. The distance of a crater or mountain from the terminator allows us to estimate the height of the sun above the lunar horizon, something we will need to know in order to estimate the height of the crater or mountain.

Consider the following sketch, not to scale.

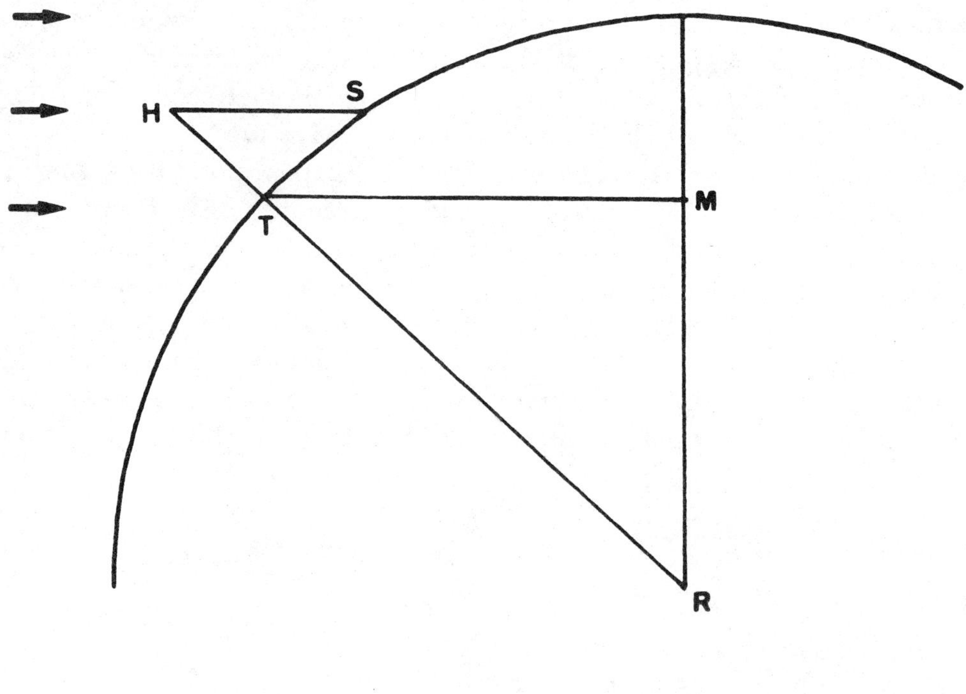

In the figure, HT = height of the mountain, HS = length of its shadow, TM = distance from the mountain to the terminator, and RT = radius of the moon. The angles THS and RTM are equal, since their sides are parallel. Further, angles HTS and TMR are both right angles. Therefore, triangles HTS and TMR are similar triangles and we can write the proportion:

$$\frac{\text{HT}}{\text{TM}} = \frac{\text{HS}}{\text{RT}}, \text{ or HT} = \frac{\text{TM}}{\text{RT}} \times \text{HS}$$

Choose a few craters in the picture. For each one, measure the length of its shadow and the distance to the terminator. Try to make your measurements as accurately as possible; you should be able to read differences as small as 0.2 mm. Also measure the radius of the moon (in centimeters. Solve the equation above for HT, the relative height of each mountain or crater.

In order to convert the relative height of the mountains to true height, you must scale your results to the true size of the moon. Look up the radius of the moon in km. Now you can solve the proportion:

$$\frac{\text{Relative radius, RT}}{\text{True radius}} = \frac{\text{Relative height, HT}}{\text{True height of mountain}}$$

How do your computed heights compare with terrestrial mountains?

Self Test

1. The earth-moon system is sometimes called a double planet because
   a. the moon's rotation period is equal to its synodic period
   b. the moon and earth are so nearly equal in size compared to the other planets and their satellites
   c. the earth is the only terrestrial planet to have a satellite
   d. the moon is geologically active, like a planet.

2. The most reasonable theory to explain the origin of the moon is
   a. the moon was captured by the earth early in the history of the solar system
   b. the earth and moon formed together
   c. the moon was once part of the earth but split off from it
   d. still under debate.

3. The length of the lunar month
   a. was originally much shorter and is gradually lengthening
   b. cannot change according to Newton's laws
   c. is gradually getting shorter
   d. oscillates with an 18.6 year period.

4. Features found only on the near side of the moon are
   a. craters
   b. maria
   c. mountains
   d. rilles.

5. Which of the following events confirms that the moon has no atmosphere?
   a. grazing occultations
   b. solar eclipses
   c. lunar eclipses
   d. the Saros cycle.

6. The distance of the moon from the earth is about
   a. 10,000 km
   b. 80,000 km
   c. 400,000 km
   d. 1 million km.

7. The lunar maria
   a. are vast plains flooded with lava
   b. are high mountains scarred by impact craters
   c. are where the oldest moon rocks were found
   d. cover the moon uniformly.

8. The oldest moon rocks are older than the most ancient rocks found on the earth because
   a. the moon was formed long before the earth
   b. the radioactive elements used for dating are rare on earth
   c. the oldest earth rocks were destroyed (recycled) by erosion and continental drift
   d. the moon formed from older material than the earth did.

9. Evidence that the moon does not have a molten core includes
   a. the lack of a lunar magnetic field
   b. seismic waves caused by quakes monitored during the Apollo landings
   c. both a and b
   d. the lack of volcanic activity on the moon.

10. Evidence that lunar craters are impact craters, rather than having been formed by volcanic eruptions includes
    a. the shape of the craters
    b. central peaks in many craters
    c. rays emanating from craters
    d. all of the above.

11. Features found only on the near side of the moon are
    a. craters
    b. maria
    c. mountains
    d. rilles.

ANSWERS TO SELECTED REVIEW PROBLEMS

2. The surface gravity of the Moon is

$$g = \frac{GM}{R^2} = \frac{(6.67 \times 10^{-8})(7.35 \times 10^{25})}{(1.74 \times 10^8)^2} = 162 \text{ cm/sec}^2.$$

If you dropped a rock from a high cliff, its velocity after 3 seconds would be $3 \times 162 = 486$ cm/sec, or about 11 miles per hour.

4. If the core has a radius of 500 km, its volume is 5.2×10^{23} cm^3. The volume of the entire Moon, which has a radius of 1738 km, is 2.2×10^{25} cm^3. Thus, only 0.02 (or 2%) of the volume of the Moon is made up of its core.

Answers to Self–Test

1–b, 2–b d, 3–a, 4–b, 5–a, 6–c, 7–a, 8–c, 9–c, 10–d, 11–b

Additional Readings

Burgess, E. 1977, The new moon — scientific results of eighteen years of lunar exploration, *Mercury*, **8**(6):10

Cadogan, P. 1983, The moon's origin, *Mercury*, **12**(2):34.

Dyal, P. and Parkin, C. W. 1971, The magnetism of the moon, *Scientific American*, **225**(2):62.

French, B. M. 1977, What's new on the moon, *Sky and Telescope*, **53**(3):164 (Part I); **53**(4):257 (Part II).

Goldreich, P. 1972, Tides and the earth-moon system, *Scientific American*, **226**(4):42.

Goles, G. G. 1971, A review of the Apollo project, *American Scientist*, **59**:326.

Hartmann, W. K. 1976, The moon's early history, *Astronomy*, **4**(9):6.

Hartmann, W. K. 1978, The early history of planet earth, *Astronomy*, **6**(8):6.

Head, J. W., Wood, C. A., and Mutch, T. A. 1977, Geologic evolution of the terrestrial planets, *American Scientist*, **65**:21.

Wetherill, G. W. 1981, The formation of the earth from planetesimals, *Scientific American*, **244**(6):162.

Wood, J. A. 1975, The moon, *Scientific American*, **233**(3):92.

Chapter 10

Venus: A Cloud-Covered Inferno

Venus, closest of all the planets to earth both physically and in general characteristics, serves as a good example of what even a moderate increase in solar radiation could do to the earth. Although otherwise similar to earth, Venus has such extremes of surface temperature and pressure that even specially designed instrument packages have survived there no more than an hour, compared with years on the moon and Mars.

The thick cloud cover on Venus prevents visual observation of the surface, but radio waves can be used to probe the planet. The reason for the high temperature, determined from radio observations, is the presence of a carbon dioxide atmosphere which traps heat via the greenhouse effect. The surface pressure is 90 earth atmospheres and the temperature is 750 K. There are three separate zones of clouds at different altitudes. These are clouds of sulfuric acid droplets rather than water vapor as on earth. The atmosphere is relatively stable, due partly to the very slow rotation of Venus.

The surface consists largely of rolling plains scarred with craters, some of which may be volcanic. A small percentage (8%) of the surface consists of highlands, akin to the continents on earth. There is evidence for tectonic activity early in the planet's history, but apparently there has not been continuing continental drift, in contrast with the earth.

With the exception of its retrograde rotation, most of the observations of Venus can be understood in the context of the theory of its formation and evolution. Its development paralleled that of earth until its crust began to cool. At that point water, formed from oxygen and hydrogen, could exist only in the form of vapor and when the atmosphere became contaminated with this gas, this triggered the greenhouse effect. The temperature became high enough to force carbon dioxide out of the surface rocks (whereas on earth this gas either stayed trapped in rocks or was recombined with rocks, with the help of reactions involving liquid water). High concentrations of carbon dioxide in the atmosphere of Venus enhanced the greenhouse effect, and the water vapor was dissociated by the resultant high temperatures. The hydrogen escaped the planet, leaving an atmosphere composed almost entirely of carbon dioxide.

Words and Phrases

Aphrodite Terra	Ishtar Terra
Beta Regis	retrograde rotation
greenhouse effect	subsolar point

Exercise: Spin-Orbit Coupling of Venus?

As you learned in Chapter 2, the sidereal period is the time it takes a planet to complete one trip around the sun with respect to the stars, and the synodic period is the time for a planet to complete one trip and return to the same configuration with respect to the earth and the sun. We can observe the synodic period directly, and from it, compute the sidereal period. As we will see, the results in the case of Venus are particularly interesting.

The synodic period of Venus can be related to the sidereal period by considering the case of two runners on a circular track. The inside runner is going faster and laps the outside runner in the equivalent of the synodic period. In the case of Venus, the synodic period is 599 days, or 1.6 earth years. The situation is illustrated in the figures below.

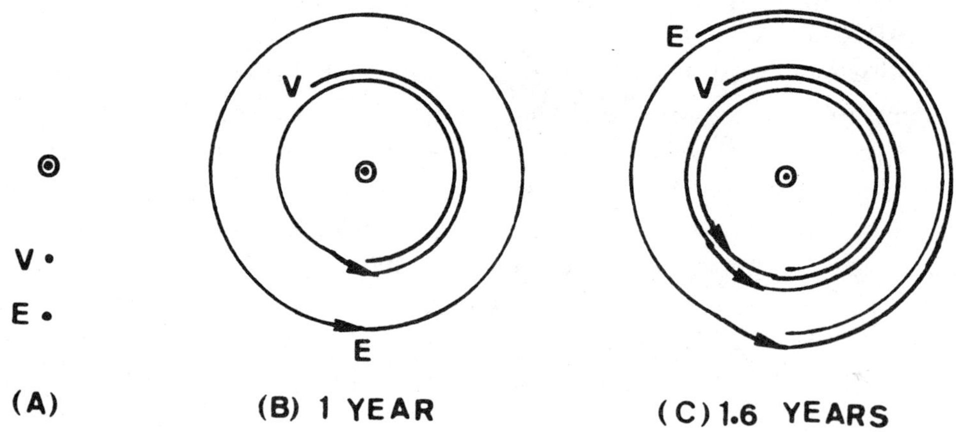

From the figure you can see that in 1.6 earth years, Venus makes 2.6 revolutions, one more than the earth does. If Venus takes 1.6 years to make 2.6 revolutions, how long does it take to make one revolution? Apply the same reasoning to Mars and determine its sidereal period if its synodic period is 2.1 years.

The relation between sidereal and synodic period in the case of Venus leads to some interesting conclusions. Suppose that some feature such as a large crater was on the side of Venus which faces the earth in figure A. We know that the rotation period of Venus is 243 days retrograde. Where will that crater be with respect to the earth in figure C? What does this suggest to you about Venus and the earth?

Self Test

1. Venus has
   a. two small satellites
   b. no satellites
   c. a ring discovered by the U.S. flyby mission
   d. one satellite.

2. In view of its dense atmosphere, it came as a surprise when soft landings on Venus revealed that
   a. sunlight reaches the surface
   b. sharp edged rocks with no signs of erosion are found on the surface
   c. shadows are seen on the surface
   d. a and b
   e. b and c.

3. Clouds on Venus are composed of
   a. water vapor
   b. carbon dioxide
   c. sulfuric acid
   d. methane.

4. Winds on Venus
   a. are strongest at high elevations
   b. are very low near the surface
   c. are responsible for giant lightning displays
   d. a and c
   e. a and b.

5. The solid surface of Venus has been mapped
    a. from the earth in visible light
    b. from spacecraft with infrared sensors
    c. from the earth and from space using radar
    d. from the earth using UV light.

6. The unusual thing about the motion of Venus is that
    a. its orbital motion is retrograde
    b. it rotates more slowly than it revolves so the day is longer than the year
    c. it rotates in the direction opposite to the other planets in the solar system
    d. none of the above.

7. The surface temperature of Venus is about
    a. 100 K
    b. 250 K
    c. 750 K
    d. 900 K.

8. Compared to the Earth, the magnetic field of Venus is
    a. weaker
    b. stronger
    c. about the same
    d. nonexistent.

9. The atmosphere of Venus is primarily composed of
    a. nitrogen
    b. water vapor
    c. methane
    d. carbon dioxide.

10. The rotation period of Venus was first determined by
    a. the Venera landings on the surface
    b. radar reflections
    c. the Mariner fly-by mission
    d. optically, with a large telescope.

ANSWERS TO SELECTED REVIEW PROBLEMS

Problems

1. Use Wien's Law, which says that $\lambda_{max} = 0.29/T$. If $T = 750\,\deg K$, then we can solve for $\lambda_{max} = 3.87 \times 10^{-4}$ cm, or 38,700 Å.

3. Since Venus has an albedo $0.76/0.39 = 1.95$ times as large as that of Earth, and receives about twice as much sunlight, Venus is $1.95 \times 2 = 3.9$ (about four) times as bright as Earth.

4. The pressure on Venus is $1300/400 = 93$ times higher than that on Earth.

Answers to Self–Test:

1–b, 2–e, 3–c, 4–e, 5–c, 6–c, 7–c, 8–d, 9–d, 10–b

Additional Readings

Head, J. W. Wood, C. A., and Mutch, T. A. 1977, Geologic evolution of the terrestrial planets, *American Scientist*, **65**:21.

Oberg, J. E. 1976, Venus, *Astronomy*, **4**(8):6.

Pettingill, G. H., Campbell, D. B., and Masursky, H. 1980, The surface of Venus, *Scientific American*, **243**(2):54.

Schubert, G. and Covey, C. 1981, The atmosphere of Venus, *Scientific American*, **245**(1):66.

Young, A. and Young, L. 1975, Venus, *Scientific American*, **233**(3):70.

Chapter 11

Mars and the Search for Life

For much of the last century Mars was believed to be the most likely place to find life, perhaps even intelligent life, outside the earth. Its similarities to our planet, known from telescopic observations, include seasonal changes in the size of the polar caps, seasonal color changes over other regions and a rotation period just over 24 hours. At the end of the nineteenth century, several observers thought they could see a network of canals, indicating a Martian civilization, and interest in Mars swelled. The "canals" did not stand the test of time however, and even before the space probes of the 1960's were launched, it was recognized that they were only an illusion. The changes in coloration were then recognized to be due to massive dust storms covering and uncovering areas of the planet, rather than seasonal variations in plant cover. The atmosphere was found, from earth-based spectroscopic measurements, to contain carbon dioxide, but no oxygen or water vapor. The temperature range was measured and found to vary from 130 K to 290 K. A Martian civilization seemed unlikely under these conditions, but primitive life was not ruled out.

The first close-up pictures of Mars were taken by *Mariner 4* in 1965. They showed a barren, crater-covered surface. In 1971 the *Mariner 9* probe was placed in orbit around Mars, and pictures sent back showed valleys, volcanos, and dried-up stream beds in addition to craters. In 1976, two U.S. *Viking* probes landed on Mars seeking evidence for life. Their tests indicated some interesting surface chemistry, but no clear evidence for anything biological. The *Viking* missions did provide a wealth of other information about Mars, however.

The Martian atmosphere is 95% carbon dioxide and 5% nitrogen, making it very similar in composition (but not pressure) to the atmosphere of Venus. Clouds of dust, as well as water and CO_2, are present in the Martian atmosphere, especially over the polar caps in late summer. Winds are typically between 35 and 50 km/hr, but since the atmosphere has only 0.006 times the pressure of earth's, their effect is much less than a comparable wind on earth. The polar caps are mostly frozen CO_2 with some water ice. There may be large reserves of water on Mars, frozen beneath the surface, and this may explain the dry channels which appear to have been formed by running water.

The surface of Mars is noted for its massive volcanic systems, far larger than any on earth. This suggests little crustal movement, which is also consistent with the lack of a Martian magnetic field, since both require a liquid core. The volcanos may also be indirectly responsible for the valleys and old river beds. If there is indeed a thick layer of permafrost beneath the surface, volcanic heating may occasionally melt it in places, and the resulting runoff could carve the observed features.

The reddish color of Mars is due to iron oxide in the surface soil and rocks. Both iron and silicon are present in high concentration, suggesting that Mars is not highly differentiated. This and other data on the average density and the apparent lack of a molten core are consistent with the conclusion that Mars was molten for only a short time and then formed a thick crust. Had Mars formed a little closer to the sun, and been warm enough for liquid water to be commonplace, perhaps it too would support life.

Words and Phrases

Deimos
laminated terrain
Olympus Mons
organic molecules
permafrost
Phobos
Syrtis Major
Valles Marineris

Exercise: The "Canals" on Mars

The "canals" that Schiaparelli observed on Mars would be quite unusual by terrestrial standards, not only for their length and coverage of the planet, but also for their width. Let us calculate the true width of the canals. Drawings of the surface of Mars showing these features generally indicated that they were between 0.5 to 4% of the diameter of the planet. Look up the diameter of Mars and compute the width of the canals in kilometers. Are there any features on earth on a scale comparable with such canals?

Remember that the canals were are hand-drawn. From your knowledge of telescopes, determine the minimum width of a feature that could have been seen through Schiaparelli's telescope. This telescope was a 9 inch refractor, so you may want to start by finding the resolution of the instrument. The formula for calculating resolution R'' in arcseconds is:

$$R'' = 2 \times 10^5 \ (\lambda/D),$$

where D is the diameter of the lens in cm and λ is the wavelength of visible light in cm.

Now use the value of R'' you have just calculated to find out the minimum width of the "canals" that Schiaparelli might have seen, assuming that their width is just equal to his ability to resolve features on Mars (be careful; was his ability to resolve features determined by the characteristics of his telescope, or of the earth's atmosphere, as described in chapter 4 of the text?). To do this, you must compare the resolution of his observations with the angular diameter of Mars, so that you can calculate the width of the "canals" by comparing them with the true diameter of the planet. Have the Mariner photographs revealed any features on Mars comparable in size to these "canals"?

Self Test

1. Which of the following is not a surface feature found on Mars
   a. large volcanoes
   b. valleys
   c. craters
   d. sand dunes
   e. none of the above.

2. The southern polar cap on mars melts almost entirely during its summer, but the northern one only shrinks during summer there. This is because
   a. the northern cap is much larger
   b. Mars is closer to the sun during the southern summer
   c. the southern cap is almost entirely CO_2 ice
   d. the winds are stronger in the north.

3. The Martian satellites Phobos and Deimos
   a. are large and spherical like the moon
   b. are small and irregularly shaped
   c. are in synchronous orbits
   d. show evidence of volcanic activity on their surface.

4. Evidence that Mars has no crustal movement (plate tectonics) comes from
   a. seasonal changes in the polar caps
   b. large shield volcanos
   c. its lack of a magnetic field
   d. more than one of the above.

5. All of the *Viking* experiments could be interpreted to show that life exists, or at least once existed, on Mars. This statement is
   a. correct
   b. wrong: organic molecules were not found, and other experiments were ambiguous
   c. wrong: all biological experiments were strongly negative
   d. wrong: only one experiment positively detected life.

6. The process in which heavy elements sink to the center of a still molten planet is called:
   a. separation
   b. reduction
   c. differentiation
   d. oxidation.

7. Surface features resembling dry river beds on Mars were probably created by
   a. ancient oceans that have evaporated
   b. lava flows
   c. underground water melted by volcanic heating
   d. melting polar caps.

8. Martian clouds are composed of
   a. blowing dust
   b. water crystals
   c. carbon dioxide crystals
   d. all of the above.

9. The reddish color of Mars is due to
   a. dust particles in the air
   b. silicon dioxide
   c. iron oxide
   d. carbon monoxide.

10. The "canals" seen on Mars by early observers
   a. were an optical illusion
   b. turned out to be chains of craters
   c. faded during heavy dust storms in recent years
   d. were confirmed by the *Mariner 9* photographs.

Answers to Selected Problems from the Text

2. Since the Earth is at 0.983 AU when it is closest to the Sun and at 1.017 AU when it is farthest away, and since the intensity of sunlight decreases as the square of the distance from the Sun, the ratio of sunlight is

$$\frac{1/(0.983)^2}{1/(1.017)^2} = 1.07.$$

This means that the intensity of sunlight at perihelion is 1.07 times as much as at aphelion for the Sun. For Mars, the intensity of sunlight is 1.45 times as high at perihelion as at aphelion, which means that the change is much more drastic on Mars than it is on Earth. This leads to larger seasonal variations.

3. The orbit of Phobos has a semimajor axis of 9.38×10^3 km (9.38×10^8 cm) and a period of 0.319 days (2.76×10^4 sec). The mass of Phobos is much smaller than the mass of Mars, so Kepler's Third Law becomes

$$M_{Mars} = \frac{4\pi^2}{G}\frac{a^3}{P^2} = \frac{4\pi^2}{G}\frac{(9.38 \times 10^8)^3}{(2.76 \times 10^4)^2} = 6.4 \times 10^{26} \text{ g}.$$

Converting to Earth masses gives the mass of Mars as .107 that of the earth, which is the value given in the text.

Answers to Self Test

1–e, 2–b, 3–b, 4–d, 5–b, 6–c, 7–c, 8–d, 9–c, 10–a

Additional Readings

Arvidson, R. E., Binder, A. B., and Jones, K. L. 1978, The surface of Mars, *Scientific American*, **238**(3):76.

Carr, M. H. 1976, The volcanoes of Mars, *Scientific American*, **234**(1):32.

Carr, M. H. 1983, The surface of Mars: a post-Viking view, *Mercury*, **12**(1):2.

Head, J. W., Wood, C. A., and Mutch, T. A. 1977, Geologic evolution of the terrestrial planets, *American Scientist*, **65**:21.

Horowitz, N. H. 1977, The search for life on Mars, *Scientific American*, **237**(5):52.

Klein, H. and Reis, R. 1977, Where are we in the search for life on Mars?, *Mercury*, **6**(2):2.

Leighton, R. B. 1970, The surface of Mars, *Scientific American*, **222**(5):26.

Leovy, C. B. 1977, The atmosphere of Mars, *Scientific American*, **237**(1):34.

Moore, P. 1980, Mars — then and now, *Mercury*, **9**(2):23.

Murray, B. C. 1973, Mars from Mariner 9, *Scientific American*, **228**(1):48.

Pollack, J. B. 1975, Mars, *Scientific American*, **233**(3):106.

Rasoul, S. I., Hunten, D. M., and Kaula, W. M. 1977, What the exploration of Mars tells us about the earth, *Physics Today*, **30**(7):23.

Veverka, J. 1977, Phobos and Deimos, *Scientific American*, **236**(2):30.

Veverka, J. and Sagan, C. 1974, McLaughlin and Mars, *American Scientist*, **62**:44.

Veverka, J., Thomas, P. and Duxbury, T. 1978, The puzzling moons of Mars, *Sky and Telescope*, **56**(3):186.

Young, R. S. 1976, Viking on Mars: a preliminary survey, *American Scientist*, **64**:620.

Chapter 12

Mercury

With Mercury, we end our exploration of the terrestrial planets. None of the other planets have earth-like conditions, even in their interiors.

Mercury is a difficult planet to study because it is always within 28° of the sun, as seen from earth. Even its rotation period was not correctly determined until the 1960's because surface features were so difficult to see. Mercury is about the size of the moon, but has a density similar to the earth. Mercury's proximity to the sun makes it very unlikely that an appreciable atmosphere could be maintained, and this was confirmed by space probes. Mercury exhibits spin-orbit coupling: it rotates in 2/3 of its period of revolution. The result is that at alternating perihelion passages it presents the same side to the sun. The reason for this is probably linked to its asymmetric mass distribution. The heavier side exhibits a huge crater basin, perhaps the scar of a collision which gave rise to its lopsided mass distribution.

Mercury's slow rotation suggests that no magnetic field should be present, yet Mariner 10 measured a weak field. The surface of the planet is lunar-like, although the craters are not closely spaced. There are prominent cliffs, called scarps, which may have resulted from a shrinking of Mercury after the crust had cooled and hardened.

Words and Phrases

Caloris Planitia
scarps
spin-orbit coupling
synchronous rotation
weird terrain

Exercise: Planetary Probes

The time that it takes a planetary probe to reach its destination is determined primarily by Kepler's third law, rather than its rocket engines. It would take a prohibitively large amount of fuel to go directly to a planet, so the probe is placed in an orbit which will intersect the planet's orbit at the correct time and place. Once the probe has escaped the earth, it is re-directed into an orbit about the sun which is calculated to intersect the orbit of the target planet when the planet is there. This is the reason the timing of the launch is critical. The shape of the orbit must be elliptical, and in the simplest case the earth represents one extreme and the planet is the other extreme of the orbit. For example, to send a probe to Mercury, one would put the probe in an elliptical orbit around the sun with aphelion at 1 A.U. (the earth's distance) and perihelion at 0.4 A.U. (the distance of Mercury from the sun). To find out how long it would take such a probe to reach Mercury we use Kepler's third law:

$$(m + M)P^2 = a^3,$$

where m represents the mass of the probe, M the mass of the sun, P the orbital period of the probe, and a the semimajor axis of its orbit. In this formula, mass is measured in solar masses, so the sun has $M = 1$ and the rocket, m, can be ignored. The semi-major axis a is measured in A.U., and is equal to one half the longest axis of the elliptical orbit. What is this axis? Now solve for the period, P. Remember that the rocket need only complete one half an orbit to get from the earth to Mercury.

Clearly a rocket may only be launched when its orbit will cause it to intersect with Mercury itself, not just the orbit of Mercury. Draw a diagram of the solar system showing the relative positions of Mercury and the Earth at the time of launch, and also at the time that the rocket arrives at Mercury.

Next apply this reasoning to Mars to find the time required to send a rocket there. Compare your answer with the launch dates and arrival dates given in Chapter 11 for Martian probes.

Self Test

1. The hottest place on Mercury is:
   a. Mare Crisium
   b. Syrtis Major
   c. Olympus Mons
   d. Caloris Planitia.

2. Scarps on Mercury are probably due to:
   a. volcanic activity
   b. ejecta from craters
   c. crustal cooling and shrinking
   d. space debris.

3. The density of Mercury is
   a. greater than the earth's because Mercury has lost more of its volatile gasses
   b. less than the earth because Mercury is smaller
   c. not yet known because no landings on Mercury have taken place
   d. 3.4 gm/cm^3.

4. The relation between Mercury's spin and orbital period is an example of spin-orbital coupling because
   a. Mercury keeps the same face to the sun at all times
   b. the same spot on Mercury always faces the sun at perihelion
   c. the planet rotates three times in every two trips around the sun
   d. all wrong. Spin-orbit coupling involves a planet and its satellite, and Mercury has no satellites.

5. Mercury is similar to the moon except that Mercury
   a. has no prominent maria
   b. has a weaker magnetic field than the moon
   c. has a thick atmosphere
   d. has a smaller density than the moon.

Matching Exercise, Ch. 8-12

1. spin-orbit coupling e
2. subduction d
3. albedo b
4. permafrost
5. greenhouse effect
6. Olympus Mons
7. maria
8. retrograde rotation
9. escape velocity
10. differentiation

a.) The process that controls continental drift

b.) The percent of light reflected by a planet

c.) The percent of light absorbed by a planet

d.) The process in which one crustal plate is pushed under the adjoining plate

e.) A relation between the spin and orbital motion of a planet

f.) The process in which heavy elements sink to the center of a molten planet

g.) A great crater on Mercury

h.) Smooth lava flows

i.) A heating process in a planetary atmosphere

j.) Underground ice

k.) The largest known volcano in the solar system

l.) The process which causes a planet to always keep one face to the sun

m.) Rotation in a counterclockwise direction as seen from above the north pole

n.) A factor in determining whether a planet can retain a certain gas in its atmosphere

ANSWERS TO SELECTED REVIEW PROBLEMS

1. The intensity goes as one on the square of the distance, so the intensity at perihelion is $(1.5)^2 = 2.25$ times brighter than at aphelion.

3. Since $a^3/P^2 = 1$ for solar orbits, we can calculate the semimajor axis of the Mariner 10 orbit by comparison with Mercury. If $P = 2P_{Merc}$, then

$$a = (2)^{\frac{2}{3}} a_{Merc} = 1.6 \times a_{Merc}.$$

Thus the orbit of Mariner 10 has a semimajor axis which is 1.6 times the semimajor axis of Mercury's orbit.

5. The escape velocity is given by

$$v_{esc} = \sqrt{\frac{2GM}{R}},$$

so using the values for M and R given in the problem above, we find that for Mercury

$$v_{esc} = 4.25 \times 10^5 \text{ cm/sec} = 4.25 \text{ km/sec}$$

and for the Moon

$$v_{esc} = 2.38 \times 10^5 \text{ cm/sec} = 2.38 \text{ km/sec}.$$

Answers to Self-Test

1–d, 2–c, 3–a, 4–c, 5–a

Answers to Review Test

1–e l, 2–d, 3–b, 4–j, 5–i, 6–k. 7–h, 8–m, 9–n, 10–f

Additional Readings

Head, J. W. Wood, C. A., and Mutch, T. A. 1977, Geologic evolution of the terrestrial planets, *American Scientist*, **65**:21.

Burgess, E. 1975, The planet Mercury, *Astronomy*, **3**(12):20.

Hartmann, W. K. 1976, The significance of the planet Mercury, *Sky and Telescope*, **51**(5):307.

Murray, B. C. 1975, Mercury, *Scientific American*, **233**(3):58.

Chapter 13

Jupiter: Giant Among Giants

With Jupiter, we begin a study of planets which are radically different from the inner terrestrial planets. We will see, however, that many of our explanations for the conditions of the inner worlds also apply to the Jovian planets.

At more than three times the distance of Mars from the sun, Jupiter is a giant planet with no solid surface and a density only one-fourth that of the earth. Sixteen moons have been discovered so far, some of these by the *Voyager* spacecraft. A great deal of our knowledge of Jupiter comes from the *Pioneer* and *Voyager* probes. The pattern of surface bands seen from earth has been resolved into intricate cyclonic currents due partly to the rapid rotation of the planet. Spectroscopy shows the presence of compounds similar to those believed to be present in the early atmospheres of the terrestrial planets. The major constituent of Jupiter is hydrogen, with helium a strong second. Unlike the terrestrial planets, which lost their free hydrogen and helium early in their evolution, Jupiter retains essentially a solar composition. In addition to the belts and zones, there are numerous spots in the atmosphere. The largest of these, the Great Red Spot, has been observed for at least 300 years, and probably represents a type of long-lived storm system.

From our observations of the surface and our knowledge of the laws of physics, it is possible to calculate what the internal structure of Jupiter must be like. Under the atmosphere lies a region of liquid hydrogen, which towards the center of the planet is under such pressure that it becomes somewhat rigid. The central core probably contains nearly all of the planet's heavy elements, which sank during the differentiation process. The presence of liquid interior zones and the rapid rotation of the planet give rise to a strong magnetic field.

The Galilean satellites, comparable to our moon in size, show increasing geologic activity from the outermost one inward. Callisto, the farthest out from Jupiter, has an ice surface covered with craters. Ganymede shows evidence for ancient tectonic activity, which has erased some of its craters. Europa resembles a marble, cracked internally, but with a smooth surface. Io shows no impact craters, and is the only body in the solar system other than the earth yet discovered to have active volcanos.

Words and Phrases

belts
Callisto
current sheet
Europa
Galilean satellites
Ganymede
Great Red Spot
Jovian
Io
rings
synchrotron radiation
zones

Exercise: Measuring the Mass of Jupiter.

Even a small telescope will show the four moons of Jupiter that were discovered by Galileo. From a single observation it is difficult to tell them apart, but repeated observations over a period of weeks will untangle their orbits. The outermost, Callisto, is the easiest to identify, and it can be used to compute the mass of Jupiter.

The diagram shows Jupiter and its four moons as they would appear on 17 successive nights. Keeping in mind that Callisto has the largest orbit, identify this satellite on as many nights as you can. You can then estimate the period (in days).

We now want to use Kepler's third law, as modified by Newton, to find the mass of Jupiter. The law can be written

$$P^2G(m_1 + m_2) = 4\pi^2a^3 \ ,$$

where P = period, a = separation (semimajor axis of Callisto's orbit), and (m_1 + m_2) = the total mass of Jupiter plus Callisto. The term G is the gravitational constant, equal to 6.67×10^{-8} dyne cm^2 gm^{-2}. All the other units must be expressed in the same system, so the period must be in seconds and a in centimeters, and the mass will be in grams.

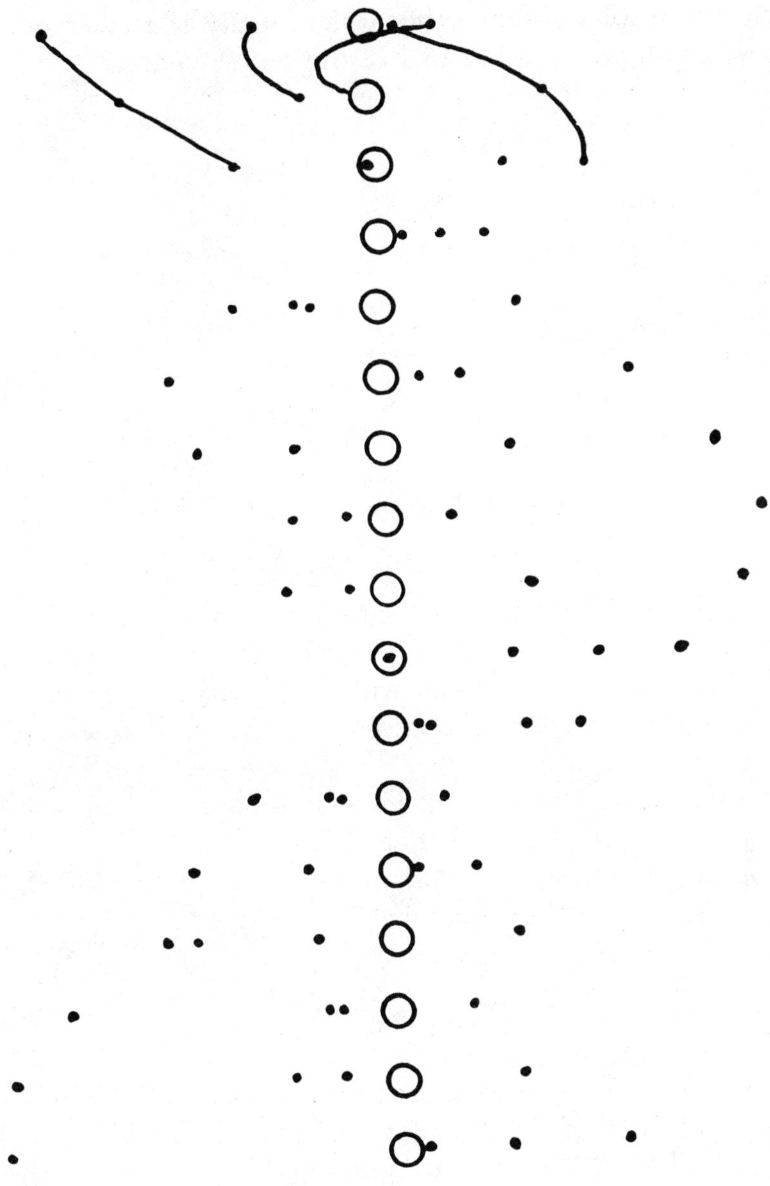

Before solving Kepler's third law, we need to find the separation a. If we measure the maximum angular separation between Jupiter and Callisto, then we can use the distance of Jupiter from the earth to compute the true separation. The answer is about 1.9×10^{11} cm.

Now solve the equation for $(m_1 + m_2)$. Clearly the mass of Callisto is much less than that of Jupiter, so your answer is really the mass of Jupiter. Compare your value with the one in table 13.1 of the text.

Self Test

1. Jupiter has a ring about it which
   a. was discovered through observations of radio synchrotron emission
   b. was discovered by the *Voyager* spacecraft
   c. lies outside the four Galilean satellites
   d. wrong: Saturn, not Jupiter, has a ring.

2. The best spot for an unmanned probe to land in order to study Jupiter would be
   a. near the Great Red spot
   b. near Jupiter's poles
   c. on Io
   d. on Europa.

3. The Galilean satellites are
   a. Phobos, Deimos, Io and Callisto
   b. Amalthea, Europa, Ganymede and Io
   c. Io, Europa, Ganymede and Callisto
   d. Ananke, Io, Europa and Ganymede.

4. The bands on Jupiter can be explained by a combination of
   a. cyclonic flow and convection
   b. differentiation and convection
   c. circulation and separation
   d. convection and rotation.

5. The mass of Jupiter is about
   a. 30 times that of the earth
   b. 100 times that of the earth
   c. 300 times that of the earth
   d. 1,000 times that of the earth.

6. Jupiter rotates differentially, which means that
   a. it rotates faster at the equator than at the poles
   b. we see only light elements on the surface since heavy elements have sunk to the core
   c. the planet is flattened at the poles
   d. it has a magnetic field which extends very far from the planet.

7. In a sense, Jupiter is more like the sun than the earth, in that
   a. its composition is mainly hydrogen and helium
   b. it is so massive that nuclear reactions occur in the core
   c. the Great Red Spot, like sunspots, is a region of intense magnetic field strength
   d. it is closer in size to the sun than to the earth.

8. The Great Spot of Jupiter is thought to be
   a. an extremely large cyclonic storm
   b. a volcano on the surface
   c. a simple cloud pattern which moves randomly
   d. a magnetic storm.

9. Jupiter is warmer than can be expected from the amount of heat it receives from the sun. This extra heat is produced by thermonuclear reactions in the core.
   a. correct
   b. wrong: heat is produced by gravitational contraction
   c. wrong: heat is produced by radioactive decay
   d. misleading: has the temperature expected from solar heating
   e. misleading: is cooler than expected from solar heating.

10. The satellite Io is interesting since it is the only body other than earth known to exhibit
   a. a mainly nitrogen atmosphere
   b. oceans with liquid water
   c. current volcanic activity
   d. primitive life.

ANSWERS TO SELECTED REVIEW PROBLEMS

Problems

2. Density is mass (M) divided by volume ($\frac{4}{3}\pi R^3$). For Jupiter's core, R = 12,000 km = 1.2×10^9 cm and M = 15 $M_\oplus$ = 9×10^{28} grams, so the density is 12.4 g/cm^3. The typical values for densities in the cores of terrestrial planets are similar to this value, but the overall density of terrestrial planets is much smaller than this.

3. Consider the Galilean satellites:

Satellite	a	P	a^3/P^2
Io	4.25×10^{10} cm	1.6×10^5 sec	3×10^{21}
Europa	6.76×10^{10} cm	3.1×10^5 sec	3×10^{21}
Ganymede	1.08×10^{11} cm	6.2×10^5 sec	3×10^{21}
Callisto	1.90×10^{11} cm	1.4×10^6 sec	3×10^{21}

If you chose units of Jupiter masses for mass, and the semimajor axis and period of the orbit of one of the satellites as units for a and P, then a^3 would equal P^2.

4. The density of Io is 3.5 g/cm^3. If its volume were doubled by adding ice (at 1 g/cm^3), then the density would be

$$\frac{1}{2}(3.5) + \frac{1}{2}(1) = 2.3 \text{ g/cm}^3.$$

This value is similar to the density of the other satellites.

Answers to Self Test:

1–b, 2–c, 3–c, 4–d, 5–c, 6–a, 7–a, 8–a, 9–b, 10–c

Additional Readings

Allen, D. A. 1983, Infrared views of the giant planets, *Sky and Telescope*, **65**(2):110.

Cruikshank, D. P. and Morrison, D. 1976, The Galilean satellites of Jupiter, *Scientific American*, **234**(5):108.

Hartmann, W. K. 1980, Moons of the outer solar system become real, although weird, places, *Smithsonian*, **10**(10):36.

Ingersoll, A. P. 1976, The meteorology of Jupiter, *Scientific American*, **234**(3):46.

Ingersoll, A. P. 1981, Jupiter and Saturn, *Scientific American*, **245**(6):90.

Millis, R. 1974, The Galilean satellites of Jupiter, *Mercury*, **3**(1):3.

Morrison, D. 1980, Four new worlds — the Voyager exploration of Jupiter's satellites, *Mercury*, **9**(3):53.

Smoluchowski, R. 1975, Jupiter 1975, *American Scientist*, **63**:638.

Soderblom, L. A. 1980, The Galilean moons of Jupiter, *Scientific American*, **242**(1):68.

Squyres, S. W. 1983, Ganymede and Callisto, *American Scientist*, **71**(1):56.

Wolfe, J. H. 1975, Jupiter, *Scientific American*, **233**(3):118.

Zimmermann R. E. 1980, Jupiter: always a starsmaid, never a star, *Griffith Observer*, **44**(4):11.

Chapter 14

Saturn and its Attendants

To anyone who has ever seen it through even a modest sized telescope, Saturn imparts a feeling of wonder and delight. The magnificence of Saturn's ring system only increases when we examine close-up pictures returned by space probes. Although it now appears that ring systems around the giant planets may be more the rule than the exception, the ring system of Saturn is the brightest, probably the most complex, and certainly the best-observed.

Saturn is similar to Jupiter in size, rotation, composition, atmospheric effects, temperature, magnetic field and radiation belts. The slightly lower temperature on Saturn has allowed the formation of more complex molecules, and has also permitted the cloud layer to exist to greater heights than on Jupiter. As a result, the belts and zones on Saturn are more obscured and muted in appearance. The belts extend much closer to the poles than they do on Jupiter, however, probably due to the seasonal variation in sunlight on Saturn (Jupiter's rotation axis is scarcely tilted, so Jupiter, unlike Saturn, does not have significant seasonal variations in solar energy input).

The current count for satellites of Saturn is 17, of which 5 were discovered by the *Voyager* missions. Several of these share orbits with others, something not discovered anywhere else in the solar system. The small inner satellites play a role in controlling the structure of the ring system, while the major satellites are interesting for other reasons. Titan is one of two satellites in the solar system known to have an atmosphere. It is enveloped in a thick shroud of methane and nitrogen, which poses some interesting questions about its formation. Iapetus has a bright and a dark side. Several of the intermediate-size moons have strange, sinuous valley systems. Enceladus has an albedo of nearly one, making it the shiniest object in the solar system. Several of the moons appear to be made of almost pure ice.

Saturn's ring system consist of particles a few meters in diameter to those microscopic in size. They are all within the Roche limit, meaning that no satellite could have formed that close to the planet, because of tidal forces.

Some of the gaps in the ring system can be attributed to the gravitational forces of the satellites. *Voyager 2* revealed far more structure in the rings than is visible from earth, making the ring system appear to be grooved like a phonograph record. The complex smaller structures are probably due to a variety of causes, ranging from small "shepherd" satellites to Saturn's magnetosphere. The magnetic field is almost certainly responsible for the spoke-like structures, seen in *Voyager* photographs, which are apparently caused by very small particles suspended out of the plane of the rings.

Words and Phrases:

Cassini division
metallic hydrogen
orbital resonance
Roche limit
shepherd satellites
spokes (in the rings)
Titan

Exercise: Saturn's Rings

The nature of Saturn's rings was understood long before the *Voyager* fly-bys. Evidence that they consist of many small chunks of material, each in orbit about the planet, came from a variety of observations. The fact that stars could be seen through the ring system argued that the rings were not solid. Furthermore, a solid ring would be shattered by differential gravitational forces. It can be shown theoretically that a satellite with the same density as Saturn would be broken into small fragments by Saturn's tidal force, if the distance of the satellite from the center of the planet were less than 2.4 times the planet's radius. From Figure 14.3, measure the distance of the ring system from Saturn in units of Saturnian radii. How close to the planet is the outer edge of the system? How close is the inner edge?

Spectroscopic studies prove that the rings are composed of individual fragments in Keplerian orbits. Remember that the Doppler shift in a spectral line tells us the velocity of approach or recession. Consider the sketch of Saturn below and answer the following questions about the Doppler shift that you would expect to observe at the labeled points.

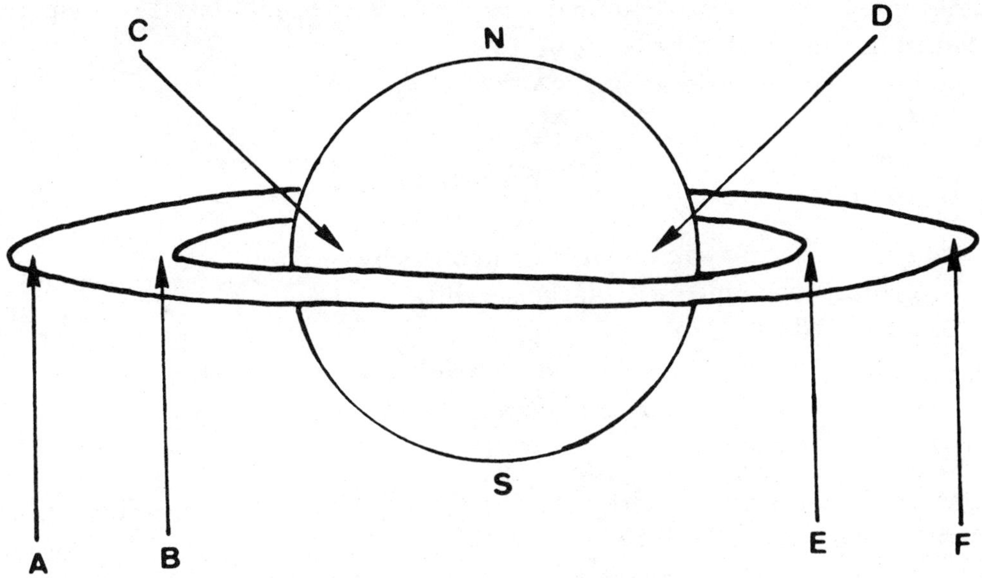

a. Suppose Saturn and its ring were not rotating at all. What Doppler shift would you measure at each point?
b. Suppose Saturn was rotating prograde (counterclockwise as seen from the north pole) and the rings were not rotating. What shifts would you see?
c. Suppose Saturn and its ring were rotating prograde, but the ring was a solid disk, like a phonograph record. What Doppler shift would you record?
d. Finally, suppose that the rings are indeed particles in Keplerian orbits. The inner ones will then be moving faster than the outer ones. What Doppler shifts will you see?

Self Test

1. The bands and zones on Saturn are not as pronounced as on Jupiter because
    a. Saturn is farther away from the earth
    b. Saturn has a thicker cloud layer overlying the bands
    c. Saturn is colder
    d. Differentiation of the elements is still going on.

2. Saturn's rings were first seen by
    a. early Greek observers
    b. Copernicus
    c. Galileo
    d. Huygens.

3. The ring structure of Saturn is controlled by
    a. orbital resonance with large satellites
    b. shepherd satellites
    c. spiral density waves caused by gravitational disturbances
    d. possible action of Saturn's magnetosphere on charged ring particles
    e. all of the above.

4. Saturn radiates more energy than it receives from the sun. The explanation is that
    a. nuclear reactions are going on in the core
    b. Saturn still retains primordial heat from its formation
    c. differentiation is still going on in the interior
    d. none of the above. Jupiter, but not Saturn, radiates more energy than it receives.

5. Because the ring system of Saturn is tilted with respect to the ecliptic, we see the rings almost exactly edge on every
    a. 6 months
    b. year
    c. 15 years
    d. 30 years
    e. never.

6. Saturn experiences seasons while Jupiter does not because
   a. Saturn has a significant axial tilt and Jupiter does not
   b. Saturn's orbit is far less circular than Jupiter's
   c. Saturn is rotating faster than Jupiter
   d. wrong. Neither Jupiter nor Saturn experience seasons.

7. Saturn is not perfectly spherical, but bulges at its equator because
   a. the gravitational attraction of the rings distorts the upper atmosphere, which is what we see
   b. limb darkening is greater in the polar regions, so the image we observe is larger at the equator
   c. the planet is rotating rapidly, causing the oblate shape
   d. the sun exerts tidal forces on Saturn.

8. The particles in Saturn's rings come from
   a. debris left over when the planet formed that was unable to coalesce into one or more moons
   b. part of a satellite broken up by a collision
   c. a comet or asteroid broken apart by tidal forces
   d. either b or c.

9. One of the largest planetary satellites, and one of the few that have atmospheres, is
   a. Phobos, which orbits Mars
   b. Titan, a moon of Saturn
   c. the earth's moon
   d. Io, a moon of Jupiter.

10. Cassini's Division is
    a. a black marking in the ring of Jupiter
    b. an Italian Army detachment
    c. a region of Saturn's rings containing material of low reflectivity
    d. a gap in the sequence of Saturnian satellites.

ANSWERS TO SELECTED REVIEW PROBLEMS

Problems

2. The rotation period of Saturn is 3.7×10^4 seconds. The radius of Saturn is 5.7×10^9 cm, so its circumference is $2\pi R = 3.6 \times 10^{10}$ cm. The velocity of a point on the surface is given by

$$\frac{3.6 \times 10^{10}\ \text{cm}}{3.7 \times 10^4\ \text{sec}} = 9.7 \times 10^5\ \text{cm/sec} = 9.7\ \text{km/sec}.$$

   This is almost 20 times faster than the velocity of atmospheric circulation (0.5 km/sec).

4. Saturn is 1.8 times as far from the Sun, so it receives only $1/(1.8)^2 = 0.3$ times as much sunlight as Jupiter. Thus, Saturn emits $2.8 \times 0.3 = 0.8$ times the energy that Jupiter receives, which is only about half as much as Jupiter emits.

6. We can use Kepler's Law in the following form:

$$\frac{a^3_{\text{Mimas}}}{P^2_{\text{Mimas}}} = \frac{a^3_{\text{part}}}{P^2_{\text{part}}}.$$

   If $P_{\text{part}} = \frac{1}{3}P_{\text{Mimas}}$, then we can solve for a_{part}. This gives $a_{\text{part}} = 0.5a_{\text{Mimas}}$, so there should be a gap in the rings at about 1.6 R_{Saturn}. This gap does exist. For Titan, using $P_{\text{part}} = \frac{1}{2}P_{\text{Titan}}$ gives $a_{\text{part}} = 0.6a_{\text{Titan}}$, so there should be a gap at about 13.4 R_{Saturn}. There is also a small gap in the rings at this radius.

Answers to Self-Test:

1–b, 2–c, 3–e, 4–c, 5–c, 6–a, 7–c, 8–a, 9–b, 10–c

Additional Readings

Allen, D. A. 1983, Infrared views of the giant planets, *Sky and Telescope*, **65**(2):110.

Hartmann, W. K. 1980, Moons of the outer solar system become real, although weird, places, *Smithsonian*, **10**(10):36.

Hunten, D. M. 1975, Saturn, *Scientific American*, **233**(3):130.
Ingersoll, A. P. 1981, Jupiter and Saturn, *Scientific American*, **145**(6):90.
Morrison, D. 1981, The new Saturn system, *Mercury*, **10**(6):162.
Owen, T. 1982, Titan, *Scientific American*, **246**(2):98.
Pollack, J. B. 1978, The rings of Saturn, *American Scientist*, **66**:30.
Pollack, J. B. and Cuzzi, J. N. 1981, Rings in the solar system, *Scientific American*, **145**(5):104.
Soderblom, L. A. and Johnson, T. V. 1982, The moons of Saturn, *Scientific American*, **246**(1):100.

Chapter 15

The Outer Planets

Far less is known about the three outer planets, Uranus, Neptune and Pluto, than about any of the others. Only Uranus has been visited by the *Voyager 2* spacecraft, which tripled the number of known satellites as well as making numerous discoveries about the magnetic field and the rings of this planet. None of the outer three planets was known before the first use of the telescope for studying the skies. Following the accidental discovery of Uranus and subsequent observations of anomalies in its motion, the existence of Neptune was deduced from the application of Newton's laws. Although it appears that the same method was applied to discover Pluto, in fact this was a lucky discovery, because any perturbation in the orbit of Neptune due to Pluto's gravitational force was far too small to have been detected.

Both Uranus and Neptune have multiple satellites, and their masses have therefore been found from Kepler's third law. They are both giant, cold spheres composed mainly of hydrogen and helium with traces of methane, very similar to Jupiter and Saturn. Uranus has a ring like Saturn's, although the albedo of the particles in the ring is very low, so the ring is difficult to observe directly. Uranus is tilted on its axis of rotation by 98° (in contrast to all the other planets, whose tilts are less than 30°). Neptune is even less well known, although it too has an enigma, in the revolution of its satellites. Triton moves retrograde and Nereid has a highly eccentric orbit.

By 1989 Pluto will be the only planet in the solar system that has not been visited by a space probe. Since it appears as a point of light in even the largest telescopes, we know almost nothing about its surface. Recently it was discovered to have a satellite, which is scarcely resolved in even the best photographs of Pluto. This led to a mass estimate, and the conclusion that Pluto is an icy body similar to the outer Galilean satellites.

Words and Phrases

Charon	Oberon
Herschel	Pluto
Miranda	Titania
Neptune	Triton
Nereid	Uranus

Exercise: A Scale Model of the Solar System

Distances within the solar system are so great that they are difficult to understand. It is therefore quite useful to consider a scale model of the sun and planets. In this exercise you will construct such a model, scaling both the size of the planets and their distance from the sun to more comprehensible values. The following table will act as a guide. First fill in the true values for the diameter and distance from the sun for each planet. You can scale them in many ways, but as an example let 1 A.U. (1.5×10^8 km) equal 100 meters. The diameter of the sun is 1.4×10^6km, which is about 1 percent of 1 A.U., so the scale model of the sun should be 1 percent of 100 meters, or 1 meter. On this scale the sun is a large beach ball and the earth is about 1 football field away. Now, using the scale 1 A.U. = 100 meters and 1.4×10^6km = 1 meter, find the scaled sizes and distances of the planets.

	Diameter	Scaled Diameter	Distance	Scaled Distance
Sun				
Mercury				
Venus				
Earth				
Mars				
Jupiter				
Saturn				
Uranus				
Neptune				
Pluto				

Self Test

1. The planet with the most extreme seasons is
   a. Mercury
   b. Earth
   c. Uranus
   d. Pluto.

2. It was once thought that Pluto might be an escaped Moon of Neptune because
   a. it has the same chemical composition as the other moons of Neptune
   b. early observers reported seeing more moons than are now present around Neptune
   c. its orbit is very eccentric, and doesn't fit the pattern of the other planets
   d. its satellite, Charon, has a period equal to Pluto's rotation period.

3. We know that Uranus has a very highly tilted rotational axis because
   a. we view the bands on the surface at a very high angle
   b. spectroscopic measurements indicate such a tilt
   c. the plane of its satellites and rings occurs at this high angle
   d. data from *Voyager 2* indicated a very high angle.

4. Although Uranus was included on star charts before Herschel observed it, it was not recognized as a planet because
   a. its motion was too slow to be noticeable
   b. it was too faint to be seen without a telescope
   c. it looked more like a star than a planet
   d. all of the above.

5. Since its discovery, Uranus has made approximately _____trips around the sun
   a. one-half
   b. two
   c. eight
   d. twenty.

6. The rings of Uranus are difficult to see because
   a. Uranus is so far from the sun
   b. the rings have a low albedo
   c. the rings are thinner than those of Saturn
   d. all of the above.

7. Neptune's satellites are unusual because
   a. Triton, which is the only satellite to have an atmosphere, moves in a retrograde orbit
   b. Triton's orbit is tilted by about 20° to the ecliptic and it moves retrograde
   c. Nereid has a long period and a circular orbit
   d. Nereid, the smaller satellite, moves in an elliptical orbit which is inclined by 28° to the ecliptic
   e. both a and c
   f. both b and d.

8. Surface markings on Uranus
   a. show cloud structure similar to Jupiter and Saturn
   b. have never been clearly observed
   c. show light and dark regions, similar to those on Mars
   d. led to the discovery of the planet's rotation.

9. Neptune was discovered
   a. by analyzing the deviations from an elliptical orbit shown by Uranus
   b. accidently, while searching for comets
   c. independently by astronomers in the U.S. and Great Britain
   d. by Clyde Tombaugh.

10. It is unlikely that Pluto could have much of an atmosphere because
   a. it is so cold that most gases would have frozen out
   b. its gravity is so small that most gases would escape
   c. both of these
   d. neither of these.

Answers to Selected Problems in the Text

2. The net force acting on Uranus can be calculated from the law of gravitation, $F = Gm_1m_2/R^2$. This gives

$$F_{total} = \frac{Gm_u}{R^2}(M_\odot - \frac{M_{Nept}}{(\frac{1}{2})^2}) = \frac{Gm_u}{R^2}(M_\odot - 4M_{Nept}).$$

If $M_\odot$ = 20,000 M_{Nept}, then this implies that the force on Uranus is reduced by 4/20000 = 1/5000 = 0.02%.

3. The sum of the masses of Neptune and Nereid is approximately the same as the mass of Neptune. Then with $M_{Nept} = 1.0 \times 10^{29}$ grams and $P_{Nereid} = 325.2$ days $= 3.2 \times 10^7$ seconds, we find

$$a^3 = P^2(\frac{GM_{Nept}}{4\pi^2}) = 5.6 \times 10^{11} \text{ cm.}$$

This is fairly typical of the sizes of orbits of the satellites of the major planets.

Answers to Self-Test

1–c, 2–c, 3–c, 4–d, 5–b, 6–d, 7–f, 8–b, 9–a, 10–c

Additional Readings

Allen, D. A. 1983, Infrared views of the giant planets, *Sky and Telescope*, **65**(2):110.

Belton, M. J. S. 1977, Uranus and Neptune, *Astronomy*, **5**(2):6

Bennett, J. A. 1981, The discovery of Uranus, *Sky and Telescope*, **61**(3):188.

Croswell, K. 1981, Triton: destined to remain a mystery, *Griffith Observer*, **45**(5):11.

Drake, S. and Kowal, C. T. 1980, Galileo's sighting of Neptune, *Scientific American*, **243**(6):74.

Elliott, J. L. Dunham, E., and Millis, R. L. 1977, Discovery of the rings of Uranus, *Sky and Telescope*, **53**(6):412.

Harrington, R. and Harrington, B. 1979, The discovery of Pluto's moon, *Mercury*, **8**(1):1.

Harrington, R. S. and Harrington, B. J. 1980, Pluto: still an enigma after 50 years, *Sky and Telescope*, **59**(6):452.

Hunten, D. M. 1975, The outer planets, *Scientific American*, **233**(3):130.

Mammana, D. L. 1978, Now we have two ringed planets in the solar system, *Smithsonian*, **9**(9):66.

Newburn, R. L. 1977, Pluto, *Astronomy*, **5**(2):10.

Ridpath, I. 1978, Pluto — planet or imposter?, *Astronomy*, **6**(12):6.

Tombaugh, C. 1979, The search for the ninth planet, *Mercury*, **8**(1):4.

Chapter 16
Space Debris

Having completed a survey of the nine major planets, we now examine the smaller bodies and the gas and dust present in the solar system. Among the objects discussed in this chapter are the comets, observed since antiquity but not recognized as extraterrestrial until studied by Tycho Brahe in the sixteenth century; meteors (or shooting stars), which were not recognized as being due to debris from space until the nineteenth century; asteroids; and interplanetary gas and dust.

Attempts to understand the spacing of the planets led to the discovery of a numerical relation which works fairly well as far out from the sun as Uranus. This relation is known as Bode's law although it is not a scientific law nor was it discovered by Bode. Since it does predict the presence of a planet between Mars and Jupiter, astronomers began searching for one late in the eighteenth century. They found instead a number of small bodies known today as the minor planets, or asteroids. The total number of asteroids may be as high as 100,000, but their total mass is much less than that of a planet. From spectroscopic analysis we know that the compositions of asteroids is highly variable, some being composed of metallic compounds while others are nearly metal-free. The accepted explanation is that the asteroids are composed of ancient material that condensed into solid form early in the history of the solar system, but which could never coalesce to form a planet due to Jupiter's gravitational influence.

Generally a comet bright enough to be seen without a telescope appears every few years, although bright city lights render many comets invisible to most people. A comet is usually visible for several weeks or months, its motion from night to night easily noticed as it passes perihelion. The most distinctive feature is a long tail, formed by gas and dust driven off the comet's head by heating from the sun. Comets are thought to originate in a very distant cloud of icy debris surrounding the sun. Occasionally a piece of this material is disturbed by gravitational forces or a near-collision, and begins to fall inward toward the sun. If it is further perturbed by the gravitational influence of one of the giant planets, its orbit may be altered in such a way that it becomes a periodic comet, passing through the inner solar system at intervals of only a few years.

A comet that regularly passes near the sun gradually loses volatile gases, until eventually nothing is left but rocky debris. This debris will disperse along the former comet's orbit, and when the earth passes through it we will see a meteor shower. Such a shower, during which many meteors per hour may be seen, all coming from the same direction, is distinct from the usual situation where particles occasionally enter the atmosphere from space and appear as randomly-oriented streaks of light. Sometimes a chunk of material survives to reach the ground, where it is called a meteorite. Study of meteorites show them to be much older than rocks on the surface of the earth. They represent the first solid material to condense out of the early solar system. The early solar system was permeated with solid chunks of this type of material, whose occasional collisions with planets and satellites created the extensive cratering seen today nearly everywhere except on the earth. Erosion has obliterated most of the evidence for cratering on the earth, but there are still signs of major craters, such as the Barringer crater in Arizona.

The space between the planets is filled with dust grains typically only 5000Å (or 5×10^{-5} cm) in diameter. The density of the dust is very low, but its presence is fairly easy to detect, because it scatters sunlight. Two well-known effects are the zodiacal light and the gegenschein. There is also gas in the space between the planets, which passes through the solar system from interstellar space, creating an interplanetary wind.

Words and Phrases

asteroids
Barringer crater
carbonaceous
Ceres
coma
comets
chondrites
dust tail
fluorescence
fireball
gegenschein
Halley's comet
interplanetary dust
ion tail
Kirkwood's gaps
meteors
meteor shower
nucleus
Pallas
zodiacal light

Exercise: The Orbit of Halley's Comet

The return of Halley's comet in 1986 attracted tremendous attention worldwide, although the comet was very difficult to see because of its orbital motion relative to the earth. To understand why this was so, it is helpful to construct a diagram showing the relative position of the earth and the comet during the time of perihelion passage. Begin by drawing a circle to represent the orbit of the earth around the sun. Indicate the direction of revolution and mark 12 intervals representing the months of the year.

Perihelion passage occurred on February 9, 1986, when Halley's comet was 0.6 A.U. from the sun. Like many comets, it moves in a retrograde orbit. We also know that meteor showers are associated with this comet, including the Aquarids on May 4, and the Orionids on October 20. This tells you the dates on which the earth passes through the orbit of the comet. Use these dates, the date of perihelion passage, and the perihelion distance, to draw the earth and its orbit around the sun, and the path of Halley's comet while it is within one A.U. of the sun. Use your figure to answer the following questions.

1. The comet was first visible through binoculars in December, 1985, just before it moved within the orbit of the earth. Was it observed in the morning or evening sky? Approximately how many degrees away from the sun was it?
2. Where did the comet appear with respect to the sun as it passed perihelion?
3. If it moved away from the sun at the same speed as it approached, when did the comet move outside the earth's orbit? At what time of night could you see it in the sky when it crossed the earth's orbit?
4. Suppose you did not know if the comet was moving in a direct or retrograde orbit. How could you figure this out from the rest of the information given in this exercise? (Hint: Remember that the Aquarids radiate from the constellation Aquarius and the Orionids from Orion. Where are these constellations at the times of each shower?)

Self Test

1. Meteors are best seen
   a. when there is moonlight to illuminate them
   b. at a quiet location where their sonic shock can be heard
   c. during a meteor shower
   d. just after sunset or just before sunrise.

2. Interplanetary dust can be detected by observations of
   a. the zodiacal light
   b. the gegenschein
   c. ocean floor material
   d. all of the above.

3. Meteorites are composed of
   a. stony chondrites
   b. iron compounds
   c. stony-iron mixtures
   d. all of the above.

4. Comets are composed of
   a. gravel embedded in frozen gases
   b. carbonaceous compounds
   c. nickle-iron compounds
   d. hydrogen and helium.

5. The asteroids are composed of
   a. metallic compounds
   b. carbonaceous compounds
   c. silicon-bearing compounds
   d. all of the above.

6. In order to estimate the size of an asteroid from the amount of light it reflects, one must make an assumption about
   a. its albedo
   b. its distance from the sun
   c. its density
   d. its temperature.

7. Which solar system objects are observed outside the ecliptic plane?
   a. asteroids
   b. comets
   c. Jovian planets
   d. none of the above.

8. Bode's law proves
   a. that the asteroids were once a planet
   b. that Neptune has moved since the solar system formed
   c. that Pluto had to exist
   d. nothing.

9. The motion of a comet in the sky is most comparable in terms of speed to the motion of
   a. an artificial earth satellite
   b. a meteor
   c. the planet Mercury
   d. the planet Saturn.

10. There is evidence for heavy cratering of the terrestrial planets early in the history of the solar system. However, this is contradicted by
    a. the number of asteroids which exist at present
    b. the lack of craters on Venus
    c. the lunar maria
    d. none of the above.

ANSWERS TO SELECTED REVIEW PROBLEMS

2. Using the ratios $P^2_{Jup}/a^3_{Jup} = P^2_{ast}/a^3_{ast}$, where $P_{ast} = 0.1\ P_{Jup}$, we find that

$$a_{ast} = 0.2a_{Jup} = 1.1\ \text{AU}.$$

3. Use Wien's law ($\lambda_m = 0.29/T$), and the Stefan-Boltzmann law ($L = 4\pi R^2 \sigma T^4$). If $\lambda_m = 0.003$ cm, then

$$T = \frac{0.29}{0.003} = 96.7\ \text{deg K}.$$

Using this temperature and a luminosity of 2×10^{20} ergs/sec, we find

$$R = \sqrt{\frac{L}{4\pi\sigma T^4}} = 5.7 \times 10^7\ \text{cm} = 570\ \text{km}.$$

4. α Cen is a star of about the same mass as the Sun. When the comet is between the Sun and α Cen, it is 100,000 AU from the Sun and 200,000 AU from α Cen. Since the force of gravity is inversely proportional to the square of the distance, the comet feels 4 times as much force from the Sun as from α Cen.

Answers to Self-Test

1–c, 2–d, 3–d, 4–a, 5–d, 6–a, 7–b, 8–d, 9–c, 10–d

Additional Readings

Cameron, I. R. 1973, Meteorites and cosmic radiation, *Scientific American*, **229**(1):64.

Cassidy, W. A. and Rancitelli, L. A. 1982, Antarctic meteorites, *American Scientist*, **70**(2):156.

Chapman, C. R. 1975, The nature of the asteroids, *Scientific American*, **232**(1):24.

Feldman, P. D. 1977, The composition of comets, *American Scientist*, **65**:299.

Grossman, L. 1975, The most primitive bodies in the solar system, *Scientific American*, **232**(2):30.

Hartmann, W. K. 1975, The smaller bodies of the solar system, *Scientific American*, **233**(3):142.

Howell, J. A. 1978, Apollo to astrobleme, *Griffith Observer*, **42**(11):2.

Lawless, J. G., Folsome, C. E., and Kvenvolden, K. A. 1972, Organic matter in meteorites, *Scientific American*, **226**(6):38.

McSween, H. Y. and Stolper, E. 1980, Basaltic meteorites, *Scientific American*, **242**(6):44.

Morrison, D. 1976, Asteroids, *Astronomy*, **4**(6):6.

Morrison, D. 1977, The diameters of minor planets, *Sky and Telescope*, **53**(3):181.

O'Keefe, J. A. 1978, The tektite problem, *Scientific American*, **239**(2):98.

Olson, R. J. B. 1979, Portrait of Halley's comet, *Scientific American*, **240**(5):160.

Tatum, J. B. 1982, Halley's comet in 1986, *Mercury*, **11**(4):126.

Van Allen, J. A. 1975, Interplanetary particles and fields, *Scientific American*, **233**(3):160.

Van Flandern, T. 1982, Where do comets come from?, *Mercury*, **11**(6):189.

Wetherill, G. W. 1979, The tektite problem, *Scientific American*, **240**(3):54.
Whipple, F. L. 1974, The nature of comets, *Scientific American*, **230**(2):49.

Chapter 17

Adding it Up: Formation of the Solar System

As is true of many far-reaching theories which encompass large subjects, the modern theory of formation of the solar system began as a rather vague general idea which has gradually been expanded and reworked in detail. In the 18th century Kant and Laplace developed the nebular hypothesis, which involved the formation of the solar system from an interstellar cloud. This theory explained many of the observations, such as the co-planar, circular orbits of most solar system objects, as well as their prograde motion in all cases. One long-standing difficulty with the theory, however, was that it could not account for the slow rotation of the sun. According to the principle of conservation of angular momentum, the sun should be spinning very rapidly, yet it does not do so. The inability of the nebular hypothesis to explain why the sun should not be spinning much more rapidly than it is led scientists to consider other theories.

Among these alternative theories was one in which a passing star was hypothesized to have pulled material out of the sun. The solar material was thought then to have condensed to form planets. Since stars are spread so thinly in space, so that close encounters would be very rare, this theory made it very unlikely that more than a few stars in the galaxy would have planets. Ever since Copernicus, astronomers have tried to avoid placing the earth in a special position, so this theory made many uneasy. There were more specific objections to the theory as well, one of which was explaining how a hot gas could cool and condense to form planets, rather than expanding and dissipating. By the early twentieth century, this and other problems led astronomers to return to the so-called evolutionary theories, ones in which the solar system was thought to have evolved naturally as a by-product of the sun's formation, as in the nebular hypothesis.

In recent years, evidence has accumulated that the solar system formed from a collapsing interstellar cloud. Advances in theoretical understanding have shown how such a collapse could explain nearly all the observed characteristics of the solar system, including the sun's slow rotation. Observational evidence has been found as well, including primarily studies of dense interstellar cloud regions where newly-formed stars are embedded.

The modern picture of the formation of the solar system envisions the formation of a flattened, rotating disk as the primordial cloud collapsed. The disk then formed condensations of solid material, which eventually coalesced to form planetesimals and then planets in the inner solar system, where most volatile elements remained in gaseous form. In the outer solar system, turbulent eddies in the solar nebula formed smaller-scale disks which condensed to form the giant planets and their extensive satellite and ring systems. The low temperatures and relatively weak solar radiation in the outer reaches of the solar system allowed even highly volatile elements such as hydrogen and helium to condense and remain in solid form, explaining why the outer planets have the same composition as the sun, while the inner planets are nearly devoid of the light-weight gases. The stubborn problem of the sun's angular momentum was finally solved in the early 1960's, when early satellite instruments detected the solar wind. The outflowing gas from the sun fills interplanetary space with charged particles, and the rotating magnetic field of the sun tries to pull them along. This creates drag on the sun, slowing its rotation in a process known as magnetic braking. This mechanism was most effective in the early history of the solar system, when the density of interplanetary gas was higher than today, and over billions of years it has slowed the solar rotation to its present rate.

If our understanding of the formation of the solar system is correct, there should be many other stars with planets. At present there is no way to detect such planets. However, steps leading to the formation of stars have been observed in cloudy regions of the galaxy, and infrared observations of many nearby stars indicate that they are surrounded by rings of material which may eventually condense to form planets.

Words and Phrases

magnetic braking
planetesimals
prograde motion
refractory elements
solar nebula
solar wind

Exercise: The Angular Momentum Problem

In this chapter you learned that early attempts to develop an evolutionary theory of the solar system were hampered by the solar rotation problem: the fact that the sun rotates too slowly for a body that condensed from a large, low density gas cloud. Conservation of angular momentum requires that it spin much faster as it contracts, in analogy with the rotating ice skater who pulls in his arms to spin more rapidly. Let's compute just how severe the problem is if one does not have magnetic braking to carry away angular momentum.

We will compute what the present-day rotational velocity of the sun would be, if no angular momentum had left the system since its beginning as an interstellar cloud. In doing this, we must take into account the angular momentum that is now contained in the planets. Let us assume that the sun formed from a cloud with a diameter of 200 A.U., a mass equal to that of the sun, and a rotational velocity at its surface of 10 km/sec. Such a velocity is typical of interstellar clouds observed in our part of the galaxy. The formula for angular momentum L of a rotating sphere is

$$L = \frac{2}{5}MVR^2,$$

where M is the mass, V is the surface velocity, and R is the radius. What is the angular momentum in units determined by the values we are using?

Now we must equate this to the present angular momentum. The angular momentum of the planets must first be added up: the formula for the angular momentum of each planet is

$$L = mvr,$$

where m is the planet's mass, v is its orbital velocity, and r is its distance from the sun. We can still use solar masses, km/sec, and A.U. as the units. The following table gives the velocity for each planet. Complete the table with the distance of each planet in A.U. and its mass, using data from your text. Compute the value mvr for each planet, and add all these values to get the total angular momentum of the planets.

	Mass	Velocity (km/sec)	Radius	L = mvr
Mercury		48		
Venus		35		
Earth		30		
Mars		24		
Jupiter		13		
Saturn		10		
Uranus		7		
Neptune		5		
Pluto		5		
			Sum =	

If the total angular momentum is the same now as it was before the original cloud began to collapse, then we have the relation:

$$L(\text{protosun}) = L(\text{planets}) + L(\text{sun})$$

and we can solve for the angular momentum of the sun. As in the case of the protosun, it is equal to:

$$L(\text{sun}) = \frac{2}{5}MVR^2$$

where M is the mass, R is the radius of the sun and V is the velocity we would expect the sun to have. You can solve for V, being careful to use the same units as before.

What period of rotation does the velocity you have derived imply? Remember the relation that velocity times the period must equal the circumference of the sun. How does this period compare with the observed period of the solar rotation?

Self Test

1. What is the major problem with the theory that a passing star pulled material from the sun to form the planets?
   a. the material would not condense
   b. a star couldn't pass close enough to pull matter from the sun
   c. the sun is mainly hydrogen and helium and the planets are mainly heavier elements
   d. none of the above.

2. According to the accepted theory of the formation of the solar system
   a. the planets formed earlier than the sun
   b. the planets and sun formed at the same time
   c. the planets formed after the sun had expelled much of the solar nebula
   d. the planets formed from cometary debris.

3. If magnetic braking had not operated over the sun's lifetime,
   a. the planets would all orbit the sun more quickly
   b. the planets would be farther from the sun than they are at present
   c. the planets would rotate faster
   d. the sun would rotate faster.

4. Astronomers assume the solar system originated from a rotating cloud which collapsed into a plane. However, this theory does not explain:
   a. the rings of Uranus
   b. the rapid rotation of the Jovian planets
   c. the retrograde rotation of Venus
   d. any of the above.

5. Any theory of the origin of the solar system must explain
   a. why planetary orbits are nearly circular
   b. why the planet satellite systems resemble the solar system
   c. why Kepler's laws hold
   d. both a and b
   e. all three, a, b and c.

6. The probable method of slowing the sun's rotation and transforming the angular momentum to the planetary material was
   a. magnetic braking
   b. angular transfer
   c. viscous flow
   d. catastrophic collision
   e. none of the above.

7. The planets account for what fraction of the total solar system mass
   a. over 90 percent
   b. around 10 percent
   c. less than 1 percent
   d. not well known.

8. Planetesimals are
   a. small bodies in orbit between Mars and Jupiter
   b. material found in comet tails
   c. small bodies that were the first to condense from the solar nebula
   d. particles making up the rings of the giant planets.

9. Refractory elements are
   a. elements that split easily during nuclear fusion
   b. elements that condense easily
   c. elements that burn
   d. elements that bend light.

10. If the evolutionary theory of planetary formation is correct, which of the following statements is true of other planetary systems?
    a. they have 7 or more planets
    b. they move in prograde (direct) orbits
    c. the planets all lie in the same plane
    d. the planetary spacing follows Bode's law.

ANSWERS TO SELECTED REVIEW PROBLEMS

1. 1 $M_\odot = 1.99 \times 10^{33}$ grams. The mass of all the planets combined is 4.5×10^{30} grams. The mass of all the satellites of the planets is insignificant by comparison. From this, we see that the Sun contains 99% of all the mass in the solar system.

3. Density is mass (M) over volume ($\frac{4}{3}\pi R^3$). The mass of the sun is 2×10^{33} grams. If this mass were spread out over a volume with a radius of 100,000 AU, the density of the cloud would be $\rho = 1.4 \times 10^{-22}$ g/cm^3. If the average atom had a mass of 2×10^{-24} grams, then there would be about 71 atoms/cm^3 in the cloud.

5. The intensity varies as the inverse square of the distance, so

$$\frac{1/(270,000)^2}{1/(4.2)^2} = 2.4 \times 10^{-10}.$$

So Jupiter would appear only 2.4×10^{-10} times as bright if it were orbiting the nearest star. This is about 24 magnitudes fainter, which might be possible to detect using the largest telescopes; however, it would be very difficult due to the proximity to the star.

Answers to Self-Test

1–a, 2–b, 3–d, 4–c, 5–d, 6–a, 7–c, 8–c, 9–b, 10–c

Additional Readings

Cameron, A. G. W. 1975, The origin and evolution of the solar system, *Scientific American*, **233**(3):32.

Falk, S. W. and Schramm, D. N. 1979, Did the solar system start with a bang?, *Sky and Telescope*, **58**(1):18.

Hartmann, W. K. 1976, In the beginning, *Astronomy*, **4**(1):6.

Hartmann, W. K. 1977, Cratering in the solar system, *Scientific American*, **236**(1):84.

Head, J. W., Wood, J. A., and Mutch, T. A., 1977, Geologic evolution of the terrestrial planets, *American Scientist*, **65**:21.

Lewis, J. S. 1974, The chemistry of the solar system, *Scientific American*, **230**(3):50.

Sagan, C. 1975, The solar system, *Scientific American*, **233**(3):23.

Schramm, D. N. and Clayton, R. N. 1978, Did a supernova trigger the formation of the solar system?, *Scientific American*, **237**(4):98.

Wetherill, G. W. 1981, The formation of the earth from planetesimals, *Scientific American*, **244**(6):162.

Chapter 18

The Sun

Our sun is a typical star, with nothing remarkable about it, except of course that it provides us with our habitable living conditions and is our principal source of energy. It is unique in another human-oriented way as well: it is the only star close enough for us to examine in great detail. What we can learn about the sun tells us a great deal about other stars.

The sun is a massive sphere of gas, composed roughly of 70 percent hydrogen, 27 percent helium and 1 percent of all the other elements. The interior of the sun is in equilibrium, the gravitational pressure inward being balanced at all points by the pressure exerted outward by the hot gas of which the sun is composed. At the center of the sun, the pressure (and hence temperature) is so great that atoms, stripped to bare nuclei, are forced to collide at such speeds that nuclear reactions take place. Four hydrogen nuclei are converted to one helium nucleus in nuclear fusion reactions, and in the process some matter is converted into energy. This energy slowly oozes to the solar surface and is released.

The sun has no solid surface, but the layer where its gaseous substance becomes dense enough to be opaque is often referred to as the surface. This layer, called the photosphere, is covered with convective cells that have the appearance of boiling oatmeal. Sunspots occur at the level of the photosphere, and they demonstrate the existence of a 22-year magnetic activity cycle in the sun. Sometimes an outburst of charged particles and energy, called a solar flare, takes place. When these charged particles reach the earth, they can interfere with radio communication.

The temperature rises both above and below the photosphere. Just above the photosphere is a layer called the chromosphere, and above that is the corona, where temperatures reach a million degrees K (the minimum value in the photosphere is about 4000 K). Occasionally hot gas arcs up into the corona, creating a temporary feature known as a prominence.

There is also a continuous flow of charged particles from the sun called the solar wind. These particles leave via coronal holes, regions in the corona where magnetic field lines open into space. The solar wind is responsible for shaping the earth's radiation belts and forcing comet tails away from the sun, among other things. Clearly the sun affects our planet in even more ways than we suspected.

Words and Phrases

chromosphere	luminosity
corona	Maunder minimum
deuterium	neutrino
differential rotation	photosphere
fission reaction	proton-proton chain
fusion reaction	solar activity cycle
granulation	solar flares
hydrostatic equilibrium	solar wind
limb darkening	sunspots

Exercise: The Temperature of the Center of the Sun

We can show from mathematical calculations that the central temperature of the sun is about 15 million K. This sort of temperature is required for nuclear reactions to take place; nowhere but in the centers of stars do we find such high temperatures. In order to have some feeling for what 15 million K really means, let us suppose that we could take a sphere of material a meter in diameter from the center of the sun and place it on the surface of the moon facing the earth. We can calculate how much this sphere alone would heat the near side of the earth. We need to use the Stefan-Boltzmann relation between luminosity, radius and temperature, and remember that the energy leaving the surface of the solar material will all pass through a sphere with a radius equal to the earth-moon distance. The energy, or luminosity, leaving the surface of the sphere of solar material is

$$L = 4\pi R_S^2 \sigma T_S^4,$$

where R_S is the radius of the sphere (50 cm), σ is a constant, and T_S is the temperature (equal to the sun's central temperature). The energy passing through the sphere centered on the solar material and having a radius R_E equal to the earth-moon distance is

$$L = 4\pi R_E^2 \sigma T_E^4,$$

where T_E is the temperature we want to determine. We can set these two energy equations equal to each other and solve for T_E:

$$4\pi R_S^2 \sigma T_S^4 = 4\pi R_E^2 \sigma T_E^4.$$

Look up the required numbers and find the temperature T_E at the distance of the earth due to this sphere of solar material. What do you think it would look like on the surface of the moon? This experiment is not possible in practice. Can you say why?

Self Test

1. Which of the following is not seen in the solar photosphere?
   a. sunspots
   b. granulation
   c. spicules
   d. limb darkening.

2. When a large solar flare is seen, we can expect radio communication difficulties a few days later. The reason for this delay is that
   a. radio waves are the last form of energy to be created in a flare
   b. charged particles travel more slowly then electromagnetic radiation
   c. particles can only escape through coronal holes
   d. it takes several days for the flare to heat the ionosphere.

3. The temperature of the corona is
   a. hotter than the photosphere although the reason for this is not understood
   b. cooler than the photosphere because the corona lies outside the photosphere
   c. the same as the photosphere although its density is much lower
   d. hotter than the chromosphere but cooler than the photosphere.

4. If nuclear reactions were to suddenly stop in the center of the sun, we could feel the effect
   a. 2 seconds later
   b. 8 minutes later
   c. 25 days later
   d. 1 million years later.
5. The Maunder minimum
   a. is a point in the "butterfly diagram" where the number of sunspots is zero
   b. is the point in the photosphere where the temperature is at its lowest
   c. was a period in the 17th century when virtually no sunspots were seen on the sun
   d. refers to the low number of sunspots every 11 years.
6. Evidence that convective motion dominates the heat flow in the outer layers of the sun includes
   a. the surface granulation
   b. the temperature gradient from center to surface
   c. limb darkening
   d. the high temperature of the corona.
7. The sun has a strong magnetic field except near the sunspots, where it is weak
   a. correct
   b. wrong: the field is weak everywhere
   c. wrong: the field is strong everywhere
   d. wrong: the field is strong near sunspots and weak everywhere else.
8. The visible part of the solar photospheric spectrum shows
   a. thousands of absorption lines, many due to iron atoms
   b. emission lines indicative of extremely high temperatures
   c. that only a minority of the chemical elements known on earth are present in the sun
   d. that the sun is hotter in its outer layers than in its interior.
9. Throughout most of the sun's volume, energy is transferred by
   a. conduction
   b. convection
   c. radiation
   d. solar wind.

10. Astronomers have recently been unable to detect _____ from the sun, which has raised questions about our ideas of stellar structure.
   a. photons
   b. sunspots
   c. neutrons
   d. neutrinos
   e. x-rays.

Answers to Selected Problems from the Text

1. The energy of a photon is given by E = hf, where h is Planck's constant. In this problem, half of the energy of a γ-ray photon with $f = 10^{20}$ hz is converted into visible photons with $f = 5 \times 10^{14}$ hz. The number of visible photons, n, can be calculated from energy conservation:

$$E_\gamma = 2nE_{visible}$$

$$h(10^{20}\text{ hz}) = 2nh(5 \times 10^{14}\text{ hz})$$

$$\Longrightarrow n = 10^5 \text{ visible photons}$$

3. To calculate the wavelength of maximum emission for thermal radiation we use Wien's law:

$$\lambda_{max} = \frac{.003}{T}\text{ meter} = \frac{3 \times 10^7}{T}\text{ Å}$$

where λ_{max} is the wavelength at which more photons are emitted than any other wavelength (the "dominant" wavelength) and T is the temperature in Kelvin.

$$\text{photosphere} : \lambda_{max} = \frac{3 \times 10^7}{6500}\text{ Å} = 4600\text{ Å (visible)}$$

$$\text{chromosphere} : \lambda_{max} = \frac{3 \times 10^7}{20,000}\text{ Å} = 1500\text{ Å (far} - \text{ultraviolet)}$$

$$\text{corona} : \lambda_{max} = \frac{3 \times 10^7}{2 \times 10^6}\text{ Å} = 15\text{ Å (x} - \text{ray)}$$

5. The intensity of thermal radiation, per unit area, is proportional to the fourth power of the temperature (Stefan-Boltzmann law). If the temperature of a sunspot were one-half that of the surrounding photosphere, then its radiation would be *less* intense by a factor of 2^4, or 16.

Answers to Self-Test

1–c, 2–b, 3–a, 4–d, 5–c, 6–a, 7–d, 8–a, 9–c, 10–d

Additional Readings

Eddy, J. A. 1977, The case of the missing neutrinos, *Scientific American*, **236**(5):80.

Foukal, P. 1980, Does the sun's luminosity vary?, *Sky and Telescope*, **59**(2):111.

Gosling, J. T. and Hundhausen, A. J. 1977, Waves in the solar wind, *Scientific American*, **236**(3):36.

Learned, J. G. and Eichler, D. 1981, A deep-sea neutrino telescope, *Scientific American*, **244**(2):138.

Newkirk, G. and Frazier, K. 1982, The solar cycle, *Physics Today*, **35**(4):25.

Pallavicini, r. and Poletto, G. 1978, Is there anything new on the sun?, *Mercury*, **7**(2):23.

Parker, E. N. 1975, The sun, *Scientific American*, **233**(3):42.

Pasachoff, J. M. 1973, The solar corona, *Scientific American*, **229**(4):68.

Pasachoff, J. M. 1978, Our sun, *Astronomy* **6**(1):6.

Tenn, J. S. 1976, The search for solar neutrinos, *Griffith Observer*, **40**(8):2.

Trefil, J. S. 1978, Missing particles cast doubt on solar theories, *Smithsonian*, **8**(12):74.

Turner, M. 1978, Neutrinos: the ultimate astrophysical probe, *Mercury*, **7**(1):9.

Van Allen, J. A. 1978, Interplanetary particles and fields, *Scientific American*, **233**(3):160.

Walker, A. B. C. 1982, A golden age for solar physics, *Physics Today*, **35**(11):61.

Wallenhorst, S. G. 1982, Sunspot numbers and solar cycles, *Sky and Telescope*, **64**(3):234.

Wolfson, R. 1983, The active solar corona, *Scientific American*, **248**(2):104.

Wilkenson, P. N. and Williamson, A. 1976, Solar activity and its influence on the weather, *Griffith Observer*, **40**(1):2.

Wilson, O. C., Vaughan, A. H., and Mihalas, D. 1981, The activity cycles of stars, *Scientific American*, **244**(2):104.

Chapter 19

Stellar Observations: Positions, Magnitudes and Spectra

Although stars are only points of light, a lot of information is contained in those points. Up until the last century the only measurements that could be made were positional measurements. Attempts to detect parallax, and confirm the earth's motion about the sun, were fruitless for many years, although in the process other motions, such as the proper motions of nearby stars, were discovered. Eventually parallax motion was detected, and as the measuring and reduction techniques improved this became an important way to measure stellar distances. The brightness of a star is another important observation. In the last century the system of measuring brightness by means of magnitudes was quantified and extended from the simple system of Hipparchus. It took only the introduction of photography, and more recently, of photoelectric detectors, for magnitude measurements to become standard.

The physical principles behind spectroscopy have been discussed in Chapter 6, and it is clear that a great deal of information about the temperature, density, ionization and velocity of a star can be deduced from its spectrum. The first stellar spectra were recorded photographically long before their physical significance was understood, but early efforts to classify stellar spectra led to a scheme which was later found to be based on temperature. The letters O, B, A, F, G, K, and M define a sequence from hottest to coolest stars.

Double, or binary stars form an important class, both because they are so common and because of the information they yield. Binary stars can be detected astrometrically from their orbital motion even if only one star is detected. They can be detected spectroscopically, either from periodic doppler shifts, or merely from the presence of two sets of superimposed spectral lines. If the earth is aligned with the orbital plane of a binary, we can detect it from periodic eclipses.

There is one class of stars which periodically fade and brighten, not because of an eclipse, but because the star pulsates. These variable stars are recognized as normal stars passing through an unstable stage of their evolution. As we will learn, some of them are useful distance indicators.

Words and Phrases

- astrometric binary
- astrometry
- B magnitude
- color index
- bolometric magnitude
- eclipsing binary
- magnitudes
- parallax angle
- parsec
- proper motion
- pulsating variable star
- spectroscope
- spectroscopic binary
- V magnitude
- visual binary

Exercise: The Many Motions of a Star

We have discussed several ways in which a star may be observed to move: precession, parallax, and proper motion are the most important ones. Let us now examine the size of these motions and see what factors they depend on.

The earliest stellar motion noted by mankind was precession, which has nothing to do with the stars themselves, but is a shift of the stars with respect to our reference frame, the earth. The period of precession is 26,000 years, so the intersection of the celestial equator and ecliptic will slide completely around the equator in this time. Calculate the rate of precession at the equator, using a shift of 360° in 26,000 years. Will this rate be the same in all parts of the sky? To answer this question start by asking yourself what the rate of precession will be at the ecliptic pole. Next consider the precession halfway between the pole and the equator.

Since parallax is so small compared to precession it took almost 2000 years longer to detect it. In speaking of parallax, we describe the star shifting back and forth against the background of more distant stars. In fact this is only an accurate description of what we see when the star is exactly in the plane of the ecliptic. Draw a sketch of what we see if the nearby star is close to the ecliptic pole. As a hint, remember that as the earth moves around the sun while we watch a star near the ecliptic pole, our line of sight traces out a cone. Finally, describe the parallactic motion of a star at intermediate latitudes between the equator and pole.

While the size of the parallax depends on the distance of the star, the proper motion of the star depends on both the distance and the star's intrinsic motion. Of the three motions discussed here, only proper motion will eventually alter the appearance of the constellations. Suppose we consider the Big Dipper. The two stars in the bowl which point to the north pole (the pointer stars) are Merak and Dubhe. Merak has a proper motion of 0.″088/year, directed 71° east of north. Dubhe has a proper motion of 0.″138/year, directed 240° east of north. Calculate how long it will take Dubhe to move 5°, approximately the present separation of these two stars. Where will Merak be then?

Self Test

1. Differences among stellar spectra are primarily due to differences in
   a. luminosity
   b. radius
   c. distance
   d. chemical composition
   e. temperature.

2. A star's color index is a measure of
   a. temperature
   b. luminosity
   c. radius
   d. distance

   because
   a. the hotter the star, the more energy it emits
   b. the luminosity is related to the surface area and hence to the radius
   c. to the shape of a star's black body curve is related to its temperature
   d. distant stars appear reddened in color.

3. The method of stellar parallax is the fundamental method for determining distances to other stars. What basic fact makes these determinations possible?
   a. the earth rotates once a day
   b. the earth revolves around the sun once a year
   c. stars have proper motions
   d. some stars pulsate.

4. The proper motion of a star depends on
   a. the velocity of the star across the line of sight
   b. the distance of the star
   c. the velocity of a star along the line of sight
   d. a and b
   e. b and c.

5. Variable stars vary mainly in
   a. spectral appearance
   b. magnitude
   c. direction
   d. radial velocity.

6. If the stars Sirius and Arcturus were each observed at six month intervals, Sirius would be found to have the greater parallax because Sirius
   a. is smaller
   b. has a higher surface temperature
   c. has a smaller mass
   d. is nearer
   e. is brighter.

7. One way of determining the temperature of a star is by its
   a. color
   b. absolute magnitude
   c. parallax
   d. declination.

8. The percentage of binary stars compared to all stars is about
   a. 10%
   b. 25%
   c. 50%
   d. 75%.

9. The spectral sequence in order of decreasing temperature, with hottest stars first, is
   a. M,K,G,F,A,B,O
   b. O,B,A,Q,M,R,S
   c. O,B,A,F,G,K,M
   d. O,B,A,F,G,M,K
   e. A,B,F,G,K,M,O.

10. A third magnitude star is _____ than a fifth magnitude star
   a. two times brighter
   b. 2.5 times fainter
   c. 6.25 times brighter
   d. 6.25 times fainter.

Answers to Selected Problems from the Text

2. If the parallax angle for an object can be measured, then the distance to that object can be determined from the formula:

$$d = \frac{1}{p}$$

where d is the distance in *parsecs*, and p is the parallax angle in *arcseconds*.

$$\text{(a)} \quad d = \frac{1}{0.25} = 4 \text{ pc}$$

$$\text{(b)} \quad d = \frac{1}{0.04} = 25 \text{ pc}$$

$$\text{(c)} \quad d = \frac{1}{0.005} = 200 \text{ pc}$$

4. From the definition of the magnitude scale, the relationship between the apparent magnitudes and brightnesses of two objects can be written as:

$$m_1 - m_2 = 2.5 \log \frac{l_2}{l_1} \iff \frac{l_2}{l_1} = 10^{\frac{m_1 - m_2}{2.5}}$$

where l_1 and l_2 are the *apparent* luminosities at Earth — that is, the power per unit area (i.e., watts per square meter) received at the Earth — of the two objects.

(a) A fourth-magnitude star (subscript 2) is about 16 times as bright as a seventh-magnitude star (subscript 1):

$$\frac{l_2}{l_1} = 10^{\frac{7-4}{2.5}} = 16$$

(b) A 17^m star is 100 times as bright as a 22^m star. No calculation is needed, since a difference of 5 magnitudes is *defined* to be a factor of 100 in brightness.

(c) A -1^m star (subscript 2) is about 630 times as bright as a sixth-magnitude star (subscript 1):

$$\frac{l_2}{l_1} = 10^{\frac{6-(-1)}{2.5}} = 630$$

6. The brightest star in the sky (Sirius: m = −1) is about *63 billion* times as bright as the faintest stars detectable to modern telescopes (m = 26):

$$\frac{l_2}{l_1} = 10^{\frac{26-(-1)}{2.5}} = 6.3 \times 10^{10}$$

8. If a star with $m_v = 13$ has a bolometric correction of −2, then its bolometric magnitude is $m_{bol} = 11$. The fraction of its light emitted in the visual band can be determined from the relation between magnitudes and brightnesses:

$$\frac{l_{vis}}{l_{bol}} = 10^{\frac{11-13}{2.5}} = .16$$

About 16% of its total energy is emitted in the visual.

10. Since the velocity of star A with respect to star B is 50 $km\ s^{-1}$, and the duration of an eclipse of star B by star A is 100 hours, the diameter of star B is simply the distance that star A travels in 100 hours:

$$\text{diameter of star B} = (50 \text{ km s}^{-1})(100 \text{ hr})(3600 \text{ s hr}^{-1}) = 1.8 \times 10^7 \text{ km}$$

Answers to Self Test

1–e, 2–a–c, 3–b, 4–d, 5–b,6–d, 7–a, 8–c, 9–c, 10–c

Additional Readings

The basic observational techniques for studying stars and the derivation of fundamental stellar properties are rarely the subject of articles in popular magazines. Therefore, the best references tend to be general astronomy texts and books on telescopes. There are a few popular articles as well:

Abt, H. A. 1977, The companions of sun-like stars, *Scientific American*, **236**(4):96.

Evans, D. S., Barnes, T. G., and Lacy, C. H. 1979, Measuring diameters of stars, *Sky and Telescope*, **58**(2):130.

McAlister, H. A. 1977, Binary-star speckle interferometry, *Sky and Telescope*, **53**(5):346.

Chapter 20

Fundamental Stellar Properties and the H-R Diagram

The fundamental properties of a star are mass, luminosity, temperature and radius. Once we know these quantities for a star, we understand from laws of physics how the star works and how it is related to all other stars. Therefore the accurate determination of these properties is an important part of stellar astronomy. Direct measurements are not possible, but we shall see that a series of steps, starting with the determination of the star's distance, yield these parameters.

The most straightforward measurement is of luminosity, or the total amount of energy emitted by the star at all wavelengths. Clearly the total energy cannot be measured, but if the distance of the star is known, the fraction of the visible energy which we detect can be computed. In order to compare stellar luminosities, one must always make allowance for the effect of their different distances from the earth. The relationship between absolute magnitude, apparent magnitude and distance is very important, since the knowledge of any two of the three quantities allows one to compute the third. We find that stellar luminosities span an enormous range, from a million times as great as the sun to ten thousand times smaller.

The temperature of a star is deduced from its spectral energy distribution or its spectral line appearance. The physical principles were outlined in Chapter 6. Measurement of a star's color or spectral class determines its temperature. Stellar temperatures range from 2000 K to 50,000 K.

The luminosity and temperature of a star are correlated in a very specific way. A plot of these quantities for a large sample of stars is called a Hertzsprung-Russell diagram, or H-R diagram. Most stars lie along a diagonal band known as the main sequence, while a few stars lie above or below this strip.

From our study of their energy distribution, we find that stars emit thermal spectra, and thus the radius of a star can be found from its luminosity and temperature according to the Stefan-Boltzmann law. The validity of using this technique is confirmed by another method of determining stellar radii: the observations of eclipsing binaries. We find that the radii of main sequence stars range from 0.1 to 20 times the solar radius.

Binary stars also allow us to determine stellar masses from the application of Kepler's third law as modified by Newton. Stellar masses may vary from 0.05 to 60 solar masses, including stars not on the main sequence.

Several additional properties of stars can be deduced from the analysis of their spectral lines. These include the rotational velocity, derived from the degree to which the lines are broadened by the Doppler effect; the chemical composition, usually deduced by computing synthetic spectra for comparison with observed ones; and the magnetic field strength, inferred from the line-splitting phenomenon known as the Zeeman effect.

We are now prepared to discuss the significance of the fundamental properties of stars. It is a tribute to the human mind that we can learn so much about objects that are only points of light, even in the largest telescopes.

Words and Phrases

absolute magnitude
apparent magnitude
bolometric magnitude
distance modulus
giant stars
H-R diagram
luminosity
magnitudes
main sequence stars
parsec
spectroscopic binary
spectroscopic parallax
white dwarf star

Exercise: Binary Stars

As noted in the text, there is a great deal of information to be extracted from a binary star. Let us consider a hypothetical binary system and see what we can get from it.

We start with the observations. Let us assume that this binary is an eclipsing, spectroscopic binary (which means almost certainly that the individual stars cannot be resolved, because only in close binaries are the orbital velocities great enough to be detected via the Doppler effect). It is observed to have a period of 10 days. During this period there are two eclipses: star A passes in front of star B and B passes in front of A. Each eclipse lasts a total of 10 hours at minimum light and another 4 hours for the light to return to its pre-eclipse level. The maximum radial velocity between star A and B during the 10 day period is 186 km/sec.

Now let's see what we can learn from these observations. First determine the diameter of each star from the eclipse information. You should recognize that during minimum light the smaller star is either completely in front of or behind the larger star. With this in mind, sketch the system (as it would appear if we could see both stars) just at the start of eclipse, 4 hours later, after 5 hours, after 6 hours, and after 10 hours. Now calculate the diameters of the two stars using the relation:

$$\text{velocity} \times \text{time} = \text{distance}$$

Next find the separation of the two stars. Remember that the two stars move around a common center of mass. The larger the mass of the star, the closer it is to the common center of mass (this is analogous to balancing two children of unequal weight on a seesaw). This problem is solved in two steps. First we find the orbital distance traveled by each star from our knowledge of the star's diameter and the length of the eclipse. Second, the ratio of eclipse duration to orbital period is the same as the ratio of the diameter of the star to the orbital distance it travels:

$$\frac{\text{diameter of star A}}{\text{orbital distance of A}} = \frac{\text{eclipse, star A}}{\text{period of system}}$$

Find the orbital distance travelled by each star, and from it, compute the radius of each star's orbit (remember that circumference = $2\pi R^2$). The sum of the radii of the two orbits is the separation of the two stars.

With the separation of the two stars, and their period, you can find the sum of the masses from Kepler's third law. Having gotten this far, you can also find the individual masses. Again, the analogy with the seesaw will help. The greater the mass (weight), the closer the star is to the center of mass (balance point of the seesaw). You can write this ratio:

$$\frac{\text{mass of A}}{\text{mass of B}} = \frac{\text{orbital radius of B}}{\text{orbital radius of A}}$$

With this ratio, and a value for the sum of the masses, you can solve for the individual masses.

There is more information available from the decrease in brightness of the binary system during eclipse. What sort of things could we expect to learn?

Self Test

1. The temperature of a star determines its
   a. color
   b. radius
   c. mass
   d. luminosity.

2. The H-R diagram is
   a. a plot of stellar temperature versus apparent magnitude for a sample of stars
   b. a plot of stellar mass versus luminosity
   c. a plot of temperature versus absolute magnitude for a sample of stars
   d. a plot of apparent magnitude versus luminosity.

3. The absence of hydrogen lines in the spectra of type K and M stars indicates
   a. there is no hydrogen in the outer atmospheres of these stars
   b. the temperature is not high enough to excite hydrogen atoms to the level required for these lines to appear
   c. these stars are very old and have entered a period of helium burning
   d. all the hydrogen in these stars is ionized.

4. To determine the sum of the masses of a binary star, we need to measure the components'
   a. temperatures and periods
   b. distance from us and size of their orbits
   c. periods and size of their orbits
   d. distance from us and the orbital velocities.

5. The absolute magnitude of a star is
   a. the magnitude of the star as seen from the earth
   b. the magnitude a star would have at a distance of 10 parsecs
   c. a measure of the star's diameter at 10 parsecs
   d. the rank of a star in its constellation.

6. Which is true of an eclipsing binary?
   a. observation of the average apparent magnitude and the period allows the luminosity to be determined
   b. the light variation is not intrinsic to the stars themselves
   c. it is a useful distance indicator
   d. the light variation has nothing to do with the orientation of the star's orbit in space
   e. none of the above.

7. The most fundamental property of a star is its
   a. temperature
   b. luminosity
   c. mass
   d. diameter.

8. Suppose you observe a previously unstudied star and find its apparent magnitude. To determine its absolute magnitude, you need to find its
   a. distance
   b. color
   c. temperature
   d. brightness as seen from earth.

9. If two stars in a binary system pass alternately in front of each other as seen from earth, we have a _____binary.
   a. spectrum
   b. passing
   c. eclipsing
   d. optical
   e. spectroscopic.

10. A star of absolute magnitude +1, compared to a star of absolute magnitude +6, is:
    a. 100 times fainter
    b. 12.5 times brighter
    c. 100 times brighter
    d. 5 times brighter.

Answers to Selected Problems from the Text

1. The *absolute* magnitude of a source is the apparent magnitude that the source would have *if* it were located 10 pc distant. The apparent (m) and absolute (M) magnitudes of a source are related to its distance (d) according to:

$$\mathrm{m - M = 5\log\frac{d}{10\ pc}} \iff \mathrm{d = 10^{1+\frac{m-M}{5}}\ pc}$$

In this problem, we are solving for absolute magnitudes:

$$\mathrm{M = m - 5\log\frac{d}{10\ pc}}$$

(a) a star 1000 pc distant with m = 12.3:

$$\mathrm{M = 12.3 - 5\log\frac{1000}{10} = 2.3}$$

(b) a star 10,000 pc distant with m = 12.3:

$$\mathrm{M = 12.3 - 5\log\frac{10,000}{10} = -2.7}$$

(c) a star 10 pc distant with m = 6.8: No calculation is required; since this star 10 pc distant, its absolute magnitude is the same as its apparent magnitude.

(d) a star 1 pc distant with m = 2.1 (in fact, there are no stars this close to the Sun):

$$\mathrm{M = 2.1 - 5\log\frac{1}{10} = 8.1}$$

3. The luminosity of a star is proportional to the radius squared and the fourth power of the temperature:

$$\mathrm{L \propto R^2T^4}$$

Here, star 2 has twice the radius and four times the temperature of star 1, so it is 1024 times as luminous:

$$\frac{L_2}{L_1} = \left(\frac{R_2}{R_1}\right)^2\left(\frac{T_2}{T_1}\right)^4$$

$$\frac{L_2}{L_1} = 2^2 4^4 = 1024$$

5. A full-width of 0.9 Å implies a half-width of 0.45 Å. Assuming that the width of the line is entirely due to rotation, the rotational velocity is:

$$\frac{v}{c} = \frac{\delta\lambda}{\lambda_0} = \frac{.45\text{Å}}{4500.9\text{Å}} = 10^{-4}\ c = 30\ \text{kms}^{-1}$$

Answers to Self Test

1–a, 2–c, 3–b, 4–c, 5—b, 6–b, 7–c, 8–a, 9–c, 10–c

Additional Readings

De Vorkin, D. 1978, Steps towards the Hertzsprung-Russell diagram, *Physics Today*, **31**(3):32.

Phillip, A. G. D. and Green, L. C. 1978, Henry Norris Russell and the H-R diagram, *Sky and Telescope*, **55**(4):306.

Phillip, A. G. D. and Green, L. C. 1978, The H-R diagram as an astronomical tool, *Sky and Telescope*, **55**(5):395.

Chapter 21

Stellar Structure: What Makes a Star Run?

In the previous chapter we learned that mass is the most fundamental parameter of a star. Now we learn the reason: a star is a gas sphere in hydrostatic equilibrium, meaning that the inward force of gravity balances the outward gas pressure. The mass of the star determines these pressures, which in turn determine the internal temperature. The internal temperature sets the stellar luminosity. A few simple physical principles therefore explain the main sequence in an H-R diagram as simply a sequence of stars with different masses. Not all stars lie on the main sequence, however. To understand this, we must consider the effect of another parameter, chemical composition.

The gas of which a star is composed is primarily hydrogen. In our discussion of the sun in Chapter 18 we learned how hydrogen is converted to helium through a fusion process at the core, where the temperature and pressure are very high. All stars on the main sequence produce energy by the conversion of hydrogen to helium. Mass is converted into energy, which works its way to the surface where we can detect it. The method by which the energy leaks out depends on how quickly the temperature decreases with distance from the center. If the temperature gradient, or drop-off with distance, is very steep, energy is transported by convection, whereas a less steep gradient will produce a radiative zone.

With only a finite amount of nuclear fuel, stars clearly have finite lifetimes. We can calculate their lifetimes based on the available hydrogen in their core and the rate at which they are converting it into energy. The lifetimes range from a few million years for the most massive stars to 5,000 billion years for the least massive ones.

As the hydrogen is converted into helium, the chemical composition of the stellar interior changes, and this changes the entire internal structure of the star. One result is increased pressure, and therefore increased temperature, which can in some cases make it possible for helium to be converted into carbon. As we shall see, a sequence of reaction stages may occur, leading to the production of elements as heavy as iron in stars of great initial mass.

Much of this chapter has dealt with quantities which cannot be observed. How, then, can we speak with confidence about the center of a star? The answer lies in what astrophysicists call a stellar model. This is a mathematical representation of a star which incorporates all the relevant physical laws as well as the observed surface features. Such models involve so many computations that only large, modern computers can handle them.

Words and Phrases

CNO cycle
convection
hydrostatic equilibrium
photosphere
radiation pressure
radiative transport
stellar evolution
stellar models
stellar winds
triple-alpha reaction

Exercise: Stellar Interiors

In this chapter we discussed the interiors of stars. Since no direct observations are possible, astronomers make mathematical models of stellar interiors, which incorporate the applicable laws of physics and the necessary observations of stellar surface conditions. One important concept is that of hydrostatic equilibrium, the balance of pressure and gravity. In this exercise, we will apply this concept to the sun and see how we can derive the pressure at its center.

Pressure is defined as the force on a specified area. To express the outward force in the sun due to gas pressure, we multiply the pressure by the area of the surface:

$$\text{outward force} = P(4\pi R^2)$$

The inward force holding the sun together is gravity:

$$\text{inward force} = GM_1M_2/R^2$$

Let us consider the two masses in this equation to be:

M_1: a shell of solar material with radius R. We can express the mass of this shell as its volume times its density. The surface area of the shell is given by $4\pi R^2$ and the thickness of the shell is r, so the volume is $4\pi R^2 r$. The density is d, so the mass is $4\pi R^2 rd$.

M_2: all the mass interior to our shell, which we can call M(R).

Now let the outward force on the shell equal the weight of the shell pulling inward and show that:

$$P = GM(R)rd/R^2$$

Use this relation to estimate the pressure at the center of the sun. Rather than working inward through a series of shells as would be done in a genuine stellar model interior, we will use only one shell. In this case both r and R equal the solar radius. If you use units of centimeters and grams then the constant $G = 6.67 \times 10^{-8}$. You will also need the average density of the sun, which you can compute from the mass and volume of the sun.

Your answer for the central pressure will be in units of dynes/cm^2. This may be more meaningful if you convert it to pounds/inch^2 using the relation 1 dyne/cm^2 = 1.45×10^{-5} lb/in^2. How does the pressure inside the sun compare with sea-level atmospheric pressure on the earth? How does it compare with the atmospheric pressure at the surface of Venus?

Self Test

1. Which kind of star puts out the most energy per unit area of its surface?
   a. G star
   b. B star
   c. F star
   d. A star
   e. M star.

2. More massive stars
   a. give off more light than less massive stars
   b. have shorter lifetimes than less massive stars
   c. have different chemical compositions than less massive a star
   d. none of the above
   e. more than one of the above.

3. The reaction which changes helium into carbon is called the
   a. CNO cycle
   b. helium flash
   c. triple-alpha reaction
   d. proton-proton cycle.

4. A star with 50 times the mass of the sun will be about
   a. 50 times as luminous
   b. 2500 times as luminous
   c. 10^4 times as luminous
   d. 10^6 times as luminous.

5. The relation between the mass of a star and its lifetime is
   a. a direct one because the more massive a star, the more fuel it has to burn
   b. an inverse one because more massive stars consume their fuel much faster than low-mass stars.
   c. the lifetime is proportional to the square of the mass because of the inverse square relation
   d. an inverse proportion because nuclear reactions are less efficient at such high temperatures.

6. At the very high densities found in the centers of stars, the atoms
   a. move at very high velocities because the temperature is so high
   b. move at very low velocities because the density is so high
   c. do not move at all because they are packed so tightly
   d. are broken apart so only protons and neutrons are found as separate particles.

7. Normal chemical reactions cannot explain the amount of energy released by stars because
   a. stars are not made of normal chemical elements
   b. stars are too hot for such reactions to occur
   c. chemical reactions release too little energy to explain the amount of energy released by stars
   d. none of the above.

8. A star in hydrostatic equilibrium is in balance between the forces of
   a. gravity and mass
   b. gravity and pressure
   c. pressure and temperature
   d. temperature and nuclear reactions.

9. As a star ages, the chemical makeup of its interior changes. What happens to the chemical makeup of the outside?
   a. we have no way of finding out
   b. the chemical makeup of the outside changes very little
   c. the same changes occur on the outside as on the inside
   d. the answer depends on the mass of the star
   e. the outside becomes richer in hydrogen.

10. A compelling reason for stating that an extremely bright star like Rigel must be relatively young is that such a star
    a. is giving out enough light to illuminate its birth processes
    b. is using up fuel at a great rate and cannot exist for long
    c. is bright because of friction with other bodies in the region from which it came
    d. is pulsating.

Answers to Selected Problems from the Test

1. The Sun radiates 4×10^{33} erg/s. If a single proton- proton chain reaction produces only 10^{-5} erg, then there must be 4×10^{38} proton-proton chain fusions taking place each second in the core of the Sun:

$$\text{number reactions} = \frac{4 \times 10^{33}\ \text{erg/s}}{10^{-5}\ \text{erg/reaction}} = 4 \times 10^{38}\ \text{reactions per second}$$

   For an *O star*, with a luminosity of 10^{38} erg/s, there must be about 10^{43} fusion reactions per second (assuming that the overall efficiency of the CNO cycle is about the same as that of the proton-proton chain).

3. When all of the hydrogen in a star's core has been converted to helium, there are only one-fourth as many nuclei as there were before the reactions. In this problem we are asked to determine the change in the core density, assuming that the average distance between nuclei remains constant.
   The mass of the core essentially is unchanged, but the volume significantly is decreased. The radius, R, of the core can be expressed as R = nd, where n is the number of nuclei across a radius, and d is the average separation between nuclei. Since the number of helium nuclei (after all of the hydrogen is fused) is one-fourth of the initial number of hydrogen nuclei, the radius is reduced by a factor of 4. Volume is proportional to the cube of the radius, so that the volume is reduced by a factor of 64. Since density is mass per unit volume, and the mass is unchanged, the density of the helium core is 64 times that of the hydrogen core.

5. A mass loss rate of 1M⊙ every 500,000 years is equivalent to 10M⊙ each 5 million years. If the star has an initial mass of 10M⊙, it would lose all of its mass in 5 million years.

Answers to Self Test

1–b, 2–e, 3–c, 4–d, 5–b, 6–a, 7–c, 8–b, 9–b, 10–b

Additional Readings

Gursky, H. and van den Heuvel, E. P. J. 1975, X-ray emitting double stars, *Scientific American*, **232**(3):24.

Perry, J. R. 1975, Pulsating stars, *Scientific American*, **232**(6):66.

Snow, T. P. 1981, Dieting for stars, or how to lose 10^{25} grams per day (and feel better, too!), *Griffith Observer*, **45**(5):2.

Weymann, R. J. 1978, Stellar winds, *Scientific American*, **239**(2):34.

Wilson, O. C., Vaughan, A. H., and Mihalas, D. 1981, The activity cycles of stars, *Scientific American*, **244**(2):104.

Chapter 22

Star Clusters and Observations of Stellar Evolution

As we learned in earlier chapters, a star's fate is determined by its mass at the time it forms. Therefore we can summarize the lives of stars by describing the history of stars with different masses.

A one solar mass star is very similar to the sun. The proton-proton chain operates for about 10 billion years; then the hydrogen in the core is gone and reactions take place in a shell around the core. The outer layers begin to expand, resulting in increased luminosity and decreased temperature. The star becomes a red giant. At this stage the gas in the core of the star is degenerate and the electrons cannot be packed any tighter. When the temperature reaches a point where the triple alpha process can begin, a runaway reaction called the helium flash takes place in the core, burning a significant fraction of the core and removing the degeneracy. The star becomes hotter now, but soon uses up its helium and becomes a red giant for the second time. We observe stars at this evolutionary stage which have blown off shells of matter. The remaining core has no fuel source left, and the star shrinks to become a white dwarf, an object the size of the earth with about the mass of the sun.

Next consider a star of 5 to 10 solar masses. Its main sequence lifetime is much shorter than that of the one solar mass star, and when the hydrogen is gone, it, too, changes into a red giant (although no helium flash takes place). If the star can lose enough mass via winds or other ejection methods to get below 1.4 solar masses, it can become a white dwarf. If not, it may explode as a supernova and leave behind a neutron star.

A very massive star, of perhaps 20 solar masses, has a main sequence lifetime is only about 6 million years, after which it evolves toward the red giant region. The picture is complicated by the mass lost through a stellar wind, which may affect both the rate of evolution and the course followed by the star in the H-R diagram. Not enough mass is lost, however, to permit the star to become a white dwarf, and probably most of these stars undergo supernova explosions.

A supernova takes place when no further nuclear fuel is available. This stage is reached after iron forms in the stellar core, because iron cannot undergo further reactions without energy input. The final remnant of the star is an expanding cloud of gas, possibly with either a neutron star or a black hole embedded within it.

Words and Phrases

black hole
degenerate gas
endothermic reaction
helium flash
horizontal branch stars
neutron star
planetary nebula
supernova
white dwarf
Wolf-Rayet star

Exercise: Star Clusters — Evolution of the Pleiades

In the exercise for chapter 21, you plotted the H-R diagram for the Pleiades. Suppose we could examine this cluster many years from now and plot its H-R diagram again. What would it look like? Evolutionary models predict the lifetimes of stars based on their mass, and observations of binaries give stellar mass as a function of spectral type, or color index. Use these relations from the following tables, along with what you learned in Chapter 22 of the text, to estimate how far each star will have moved in the H-R diagram after 107 years, and again after 108 years.

Numbers from Evolutionary Models

Mass (solar masses)	Time on Main Sequence (years)	Time to Reach Red Giant Branch (years)
20	6×10^6	6×10^5
5	10^8	2×10^6
1	10^{10}	10^9

Relation between Mass, Spectral Type and (B-V)

Mass	Spectral type	B-V
17	BO	-0.3
3.6	AO	+0.0
1.8	FO	+0.3
1.1	GO	+0.6

Self Test

1. The bright binary star Sirius consists of a main sequence A-star and a white dwarf. More than 1500 years ago, astronomers reported that Sirius was a bright red star. If we believe this report, what is the explanation?
   a. at that time, Sirius A was moving away from us in its orbit, so that it appeared red
   b. at that time, Sirius A was a red giant
   c. at that time, Sirius B was a red giant
   d. at that time, Sirius A was a main sequence O star.

2. During its lifetime, the sun will pass through all of the following stages except:
   a. main sequence
   b. supergiant
   c. white dwarf
   d. red giant.

3. Which of the following stars will spend the most time on the main sequence?
   a. a 3 solar mass star
   b. a 10 solar mass star
   c. the sun
   d. a 0.5 solar mass star.

4. A star's evolutionary path in the H-R diagram could be compared to
   a. a mountain climber as he goes from the bottom to the top
   b. a person's changing height and weight as they age from 1 to 50 years
   c. the evolution of Neanderthal man to the present human race
   d. the annual migration of a bird.

5. Astronomers currently believe that elements heavier than iron are created in
   a. the initial explosion of the universe
   b. proton-proton chains
   c. degenerate white dwarfs
   d. supernova explosions.

6. What is an evolutionary track?
   a. the path of a star through space
   b. a dark filamentary feature on the surface of a star
   c. a graph of temperature at various positions inside a star
   d. the sequence of positions in the H-R diagram that the star occupies as it undergoes evolutionary changes
   e. the curve traced out around a star by its planetary nebula.

7. Astronomers on the Battlestar Galactica have discovered a star that evolves in a strange fashion: its luminosity decreases while its temperature increases. In what direction does this star "move" in the HR diagram?
   a. upper right to lower left
   b. upper left to lower right
   c. straight up and down
   d. left to right
   e. lower right to upper left.

8. The probable fate of our sun is
   a. to expand as a red giant, undergo a nova outburst and end as a white dwarf
   b. to expand as a red giant, eventually become a planetary nebula, and end as a white dwarf
   c. expand, undergo a helium flash, become variable, and eventually to explode as a supernova
   d. to become a black hole.

9. Nuclear fuel sources in stars include
   a. hydrogen
   b. hydrogen and helium
   c. hydrogen, helium, and elements as heavy as iron
   d. hydrogen, helium, carbon and uranium.

10. A star like the sun will spend the largest fraction of its life
    a. contracting from an interstellar cloud
    b. as a main sequence star
    c. as a red giant
    d. as a planetary nebula.

Answers to Selected Problems from the Text

1. Using the formulas for density (ρ), escape velocity (v_e), and surface gravity (g), we can calculate values relative to the Sun by:

$$\text{density}: \frac{\rho}{\rho_\odot} = \frac{\frac{M}{M_\odot}}{(\frac{R}{R_\odot})^3}$$

$$\text{escape velocity}: \frac{v_e}{v_{e\odot}} = \sqrt{\frac{\frac{M}{M_\odot}}{\frac{R}{R_\odot}}}$$

$$\text{surface gravity}: \frac{g}{g_\odot} = \frac{\frac{M}{M_\odot}}{(\frac{R}{R_\odot})^2}$$

   For a $1M_\odot$ red giant with a radius of $20R_\odot$: density is *decreased* by a factor of 8000; escape velocity is 0.22 of the solar value; and surface gravity is *reduced* by a factor of 400 (compared to the corresponding solar values).
   For a $1M_\odot$ white dwarf with a radius of $0.01R_\odot$: density is *increased* by a factor of 10^6; escape velocity is increased by a factor of 10; and surface gravity is *increased* by a factor of 10^4 (compared to the corresponding solar values).

3. The wavelength of maximum emission for a star with a surface temperature of 100,000 K can be calculated from Wein's law:

$$\lambda_{max} = \frac{3 \times 10^7}{T} \text{ Å}$$

   Plugging in $T = 100,000$, we find $\lambda_{max} = 300$ Å. An ultraviolet telescope would be best for observing this object, since its wavelength of maximum emission is in the extreme ultraviolet portion of the spectrum.

5. A supernova with absolute magnitude −19 is *4 billion* times brighter than the Sun (absolute magnitude +5):

$$\frac{l_2}{l_1} = 10^{\frac{m_1-m_2}{2.5}} = 10^{\frac{5-(-19)}{2.5}} = 4 \times 10^9$$

Answers to Self-Test

1–c, 2–b, 3–d, 4–b, 5–d, 6–d, 7–a, 8–b, 9–c, 10–b

Additional Readings

Flannery, B. P. 1977, Stellar evolution in double stars, *American Scientist*, **65**:737.

Green, L. C. 1977, Supernovae today, *Sky and Telescope*, **54**(1):11.

Herbst, W. and Assousa, G. E. 1979, Supernovas and star formation, *Scientific American*, **241**(2):138.

Kirschner, R. P. 1976, Supernovas in other galaxies, *Scientific American*, **235**(6):88.

Schramm, D. N. and Arnett, W. 1975, Supernovae, *Mercury*, **4**(3):16.

Schramm, D. N. and Clayton, R. N. 1978, Did a supernova trigger the formation of the solar system?, *Scientific American*, **239**(4):98.

Sweigart, A. V. 1976, The evolution of red giant stars, *Physics Today*, **29**(1):25.

Stephenson, F. R. and Clark, D. H. 1976, Historical supernovas, *Scientific American*, **234**(6):100.

Williams, R. E. 1981, The shells of novas, *Scientific American*, **244**(4):120.

Wyckoff, S. 1979, Red giants: the inside scoop, *Mercury*, **8**(1):7.

Chapter 23

Life Stories of Stars

In many scientific fields it is possible to learn about important processes, such as the growth of a living organism or a chemical reaction, by watching them take place. Events in astronomy take place so slowly, however, that all of our accumulated observations since early history still constitute only a snapshot. We must deduce the life histories of stars indirectly.

Some of the best evidence for stellar evolution comes from the study of star clusters. Clusters range from very loose groupings of several dozen stars to giant spherical groups containing tens or hundreds of thousands of stars. In every case, however, the stars in a cluster are at the same distance from us, and we can safely assume that they all formed together, at the same time and out of the same gas.

If the stars in a cluster formed together from the same material, then the major differences among them are their masses. Hence the H-R diagram for a cluster shows a main sequence representing the range of masses that are present. We learn about the aging process in stars by comparing H-R diagrams of clusters of different ages, which allows us to see how stars of different masses evolve. In some clusters, which are deduced to be old ones, there are no stars on the upper main sequence but there are red giants in the upper right-hand corner of the diagram. In other clusters, which turn out to be young ones, there are stars on the main sequence all the way up to the top, but there are no red giants. A combination of these observations with theory of stellar structure led to the conclusion that stars move to the right and become red giants as their hydrogen fuel is used up. The more massive stars evolve faster and leave the main sequence to become red giants sooner than their less massive cluster-mates, which explains why the stars on the upper part of the main sequence move away to the right first.

In the H-R diagrams of very young clusters, we find stars above the lower part of the main sequence; these are evidently newly-formed stars that have not yet reached the main sequence. Such clusters contain a great deal of gas and dust, probably the remnants of the material from which the stars formed, and often infrared-emitting objects thought to be condensations collapsing to become stars are found embedded within these clouds.

It is important to realize that the comparison of different H-R diagrams only indicates the relative ages of the stars. To obtain absolute ages, we must rely on model stellar interiors which tell us how long a star of a given mass will last. In the case of our sun, these calculations are supported by measurements of the age of the earth, derived from geologic methods.

Words and Phrases

globular cluster
H II region
horizontal branch stars
Hyades
main sequence fitting
main sequence turn-off
OB association
open cluster
Ring nebula
T Tauri star
zero age main sequence

Exercise: Star Clusters

The Pleiades are a bright open cluster whose brightest members (the "Seven Sisters") are visible without a telescope. In this exercise you will determine the distance to this cluster and derive some of its properties.

Below is a list of observations of some of the stars in the cluster. Each star has been studied photometrically and its magnitude, V, and color, B–V, have been measured. Plot a color-magnitude diagram for the cluster (this should be done on graph paper). Now use the method of main sequence fitting described in the text to find the distance modulus of the cluster. Use the standard H-R diagram shown in the text. You may find it helpful to redraw the standard H-R diagram to the same scale as your diagram of the Pleiades. What is the distance to the Pleiades?

Now answer the following questions about this cluster.

a. Which are the four most massive stars in your color-magnitude (i.e. H-R) diagram?

b. Which stars in your diagram do you think are the "Seven Sisters"?

Stars in the Pleiades*

Star Number	V	B-V	Star Number	V	B-V
3	8.2	.09	26	7.9	.04
5	8.1	.04	27	5.7	–.22
7	9.6	.33	28	6.4	–.20
8	8.1	.12	31	4.2	–.24
9	9.8	.41	32	10.4	.53
10	5.4	–.22	33	7.4	–.07
11	3.7	–.30	34	8.1	.21
13	10.4	.51	35	10.2	.62
15	8.6	.20	36	9.3	.34
16	5.6	–.26	39	10.5	.53
17	4.3	–.30	40	6.8	–.15
18	8.9	.31	41	8.4	.15
19	8.0	.07	42	9.4	.34
20	8.6	.21	43	6.9	–.14
21	10.1	.49	44	7.6	.05
22	7.2	.01	45	7.3	–.13
23	9.7	.43	46	7.7	.00
24	9.4	.37	47	6.8	–.10
25	3.9	–.25	56	6.9	–.03

* Data from Johnson, H. L. and Morgan, W. W. 1951, *Astrophysical Journal*, 114:522.

Self Test

1. The length of time it takes for a gas cloud to collapse to form a star
   a. is longer for more massive stars than for less massive stars
   b. is about one hundred thousand years, so we see few in this process
   c. is shorter the more massive the star is
   d. is not yet known.

2. If a cluster exhibits a main sequence turn-off at spectral type F2, it means that
   a. the cluster has no stars more massive than those of type F2
   b. stars more massive than F2 stars have not yet formed
   c. stars more massive than F2 stars have evolved off of the main sequence
   d. stars less massive than F2 stars have not yet evolved onto the main sequence.

3. When a new star like our sun is forming what is the event that halts the initial gravitational contraction?
   a. the onset of fusion
   b. the helium flash
   c. the core becoming degenerate
   d. the exhaustion of fuel supplies.

4. H-R diagrams have been constructed for two clusters. Cluster No. 1 contains main sequence stars from type B through type M. Cluster No. 2 contains main-sequence stars from type A through type M. Which cluster is <u>older</u>?
   a. No. 1
   b. No. 2
   c. impossible to tell without more data
   d. both clusters are the same age

5. Star clusters are useful in the study of stellar evolution because
   a. all the stars in the cluster are at the same distance and were formed at about the same time
   b. comparing the H-R diagram of one cluster with another allows you to determine the age of each cluster
   c. all the stars in the cluster have the same mass and are at the same distance from you.
   d. the stars in a cluster all evolve at the same rate.

6. To "see" inside a dense interstellar cloud, astronomers must use
   a. ultraviolet observations
   b. infrared observations
   c. spectroscopic observations
   d. model star calculations.

7. Which star clusters are associated with recent star formation@
   a. globular
   b. associations
   c. open
   d. both a and b
   e. none of the above

8. A massive, spherical cluster of old stars is
   a. an open cluster
   b. a globular cluster
   c. an association
   d. a union

9. An H II region is
   a. a cloud of hydrogen molecules
   b. an atmosphere of hydrogen around stars like our sun
   c. a source of 21-cm radiation
   d. a sphere of ionized gas surrounding a hot star
   e. a cloud in which water molecules have been found.

10. In a collapsing protostar, the increased rotation rate is due to
   a. conservation of angular momentum
   b. conservation of linear momentum
   c. conservation of energy
   d. all of the above.

Answers to Selected Problems from the Text

2. The age of a cluster with a main sequence turnoff point at F5 is the main sequence lifetime of an F5 star. According to table 20.2, an F5 star has $1.3M_\odot$ and $2.5L_\odot$. Its lifetime is:

$$\frac{t}{t_\odot} = \frac{\frac{M}{M_\odot}}{\frac{L}{L_\odot}} = \frac{1.3}{2.5} = .5$$

The age of the cluster is half of the solar lifetime, or 5 billion years. The B − V color index at the turnoff point (F5) is +.42.

3. Luminosity is proportional to the square of the radius and the fourth power of the temperature. If the Sun becomes a red giant with a radius 20 times its present value, and a temperature one-half its present value, then its luminosity will change by a factor of $20^2(\frac{1}{2})^4$: it will increase by a factor of 25.
To find the change in the bolometric magnitude, we use:

$$\mathrm{m}_1 - \mathrm{m}_2 = 2.5 \log \frac{l_2}{l_1} = 2.5 \log 25 = 3.5$$

The bolometric magnitude will *decrease* by 3.5 magnitudes.

Answers to Self Test

1–c, 2–c, 3–a, 4–b, 5–a, 6–b, 7–b, 8–b, 9–d, 10–a

Additional Readings

Bok, B. J. 1972, The birth of stars, *Scientific American*, **227**(2):48.
Bok, B. J. 1981, Early phases of star formation, *Sky and Telescope*, **61**(4):284.
Cohen, M. 1981, Are we beginning to understand T Tauri stars?, Sky and Telescope, **62**(4):300.
Cohen, M. 1981, The shocking truth about Herbig-Haro objects, *Griffith Observer*, **45**(1):9.
Dickman, R. L. 1977, Bok globules, *Scientific American*, **236**(6):66.
Iben, I. 1970, Globular cluster stars, *Scientific American*, **223**(1):26.
Lada, C J. 1982 2, Energetic outflows from young stars, *Scientific American*, **247**(1):82.
Lesh, J. R. 1978, Swarms of stars: cosmic calibrators, *Astronomy*, **6**(3):6.
Loren, R. B. and Vrba, F. J. 1979, Starmaking with colliding molecular clouds, *Sky and Telescope*, **57**(6):521.
Neugebauer, G. and Becklin, E. E. 1973, The brightest infrared sources, *Scientific American*, **228**(4):28.
Schramm, D. N. and Clayton, R. N. 1978, Did a supernova trigger the formation of the solar system?, *Scientific American*, **239**(4):98.
Seeds, Michael A. 1979, Stellar evolution, *Astronomy*, **7**(2):6.

Sweigart, A. V. 1976, The evolution of red giant stars, *Physics Today*, **29**(1):25.

Stephenson, F. R. and Clark, D. H. 1976, Historical supernovas, *Scientific American*, **234**(6):100.

Wallerstein, G. and Wolff, S. 1981, The next supernova?, *Mercury*, **10**(2):44.

Wynn-Williams, G. 1981, The newest stars in Orion, *Scientific American*, **245**(2):46.

Zeilik, M. 1978, The birth of massive stars, *Scientific American*, **238**(4):110.

Chapter 24

Stellar Remnants

The form in which matter can exist is varied indeed. We have already seen this to some extent. Starting from the everyday properties of matter on earth, we have learned of the extremes of temperature exhibited from Venus to Pluto, and we have discussed the central regions of the sun, with undescribable temperatures and pressure. Matter can exist in still more bizarre fashions, however, and perhaps the most extreme forms are those it takes in stars that have exhausted their nuclear fuel.

There are three possibilities for stellar remnants: white dwarfs, neutron stars and black holes. Which of these final forms a star will take depends on its mass when the nuclear fuel runs out. If the mass is less than 1.4 solar masses, a white dwarf will result. This is a compact, dense stellar remnant about the size of the earth, yet containing an amount of mass comparable to that of the sun. The interior of a white dwarf is a degenerate electron gas, a strange kind of gas whose pressure is created by the resistance of electrons to being squeezed together beyond a certain limit. Many white dwarfs are binaries, and in the case that the companion star is still burning its nuclear fuel, sometimes an event known as a nova takes place. The companion sheds mass, by swelling up to become a red giant or through a stellar wind, and some of its cast-off material may fall onto the white dwarf. This triggers an instant nuclear reaction outburst, and the white dwarf brightens to rival a supergiant for a brief time.

Stars too massive to become white dwarfs probably all explode as supernovae. A supernova leaves an expanding gas cloud, and possibly a stellar remnant. These gas clouds are sometimes detected in visible light, as is the well-known Crab Nebula, but often can only be observed at radio wavelengths. The fate of the stellar remnant, if there is one, depends on its mass. For stellar masses between 1.4 and about 3 solar masses, the matter collapses to form a degenerate neutron gas, and the radius of the star is about 10 km. Neutron stars are so small that they are not visible like white dwarfs.

There are circumstances, however, under which we can detect neutron stars. In one case, the star may be observed as a pulsar, because a spinning neutron star can beam energy out from its magnetic poles. If this beam crosses the earth, we detect brief, rapid flashes of radiation. The second case is that of a binary system in which matter from a normal star is lost to the neutron star. As the gas spirals in, it is compressed and heated to the point where it emits X-rays.

The most bizarre form of matter results from the collapse of a star with a mass greater than about three solar masses. In this case there is no pressure sufficient to stop the collapse, and the matter becomes concentrated in an infinitesimally small region called a singularity, or a black hole. The behavior of matter and energy under these conditions is predicted from Einstein's theory of general relativity. Light cannot escape the black hole, and time is distorted. The only hope of detecting such an object is to do so indirectly: if we find an invisible object with a large enough mass, it is probably a black hole. A few binary systems have been observed in which an unseen companion appears to be a very good candidate.

Words and Phrases

accretion disk
binary X-ray source
event horizon
gravitational redshift
nova
recurrent nova
Schwarzschild radius
singularity
supernova remnant
synchrotron emission

Exercise: Spin Rates of Stellar Remnants

There is direct observational evidence that many stellar remnants spin very rapidly. Pulsars, for example, are thought to be rapidly spinning neutron stars. Although there are ways in which a star can lose mass, and therefore angular momentum, let us compute some example rotational periods assuming that angular momentum is conserved. In an earlier exercise the angular momentum of a rotating sphere was given as $\frac{2}{5}\mathrm{mvr}^2$. If the mass, m, does not change, then a decrease in radius, r, will mean that a corresponding increase is required in surface velocity, v. The period can be found from a variation of the relation velocity $\times$ time $=$ distance:

$$\text{period} = \text{circumference/velocity} = 2\pi \mathrm{r}/\mathrm{v}.$$

Compute the velocity of each object in the following table after its collapse, and also the object's period. Use the numbers given for the original velocity and the decrease in radius following the collapse.

	White Dwarf	Neutron Star	Black Hole
Original Velocity	10 km/sec	50 km/sec	50 km/sec
Fractional Decrease	$\frac{7\times10^{3}\text{km}}{7\times10^{5}\text{km}}$	$\frac{10\text{km}}{1.2\times10^{6}\text{km}}$	$\frac{\text{(less than 1 km)}}{6\times10^{6}\text{km}}$

Self-Test

1. All novae are
   a. red stars
   b. Cepheids
   c. sites for the formation of pulsars
   d. close binaries.
2. Which event does not involve ejection of gas by a star?
   a. planetary nebula formation
   b. nova
   c. shell hydrogen burning
   d. stellar wind.
3. 1.4 solar masses is the upper limit on the masses of
   a. black holes
   b. supergiants
   c. neutron stars
   d. white dwarfs.
4. How many stars now known to exist within 10 parsecs of the sun are likely to become supernovae eventually?
   a. none
   b. 5 to 100
   c. more than 100
   d. 1.

5. The reason that visible supernova explosions are so rare is:
   a. most supernovae are too far away for us to observe them through the obscuring gas and dust
   b. a very small percentage of stars are massive enough to evolve into supernovae
   c. only close binaries can evolve the necessary iron core needed for a supernova explosion
   d. many supernova explosions emit only radio radiation, and are not visible.

6. What phenomenon provides observational evidence for the existence of neutron stars?
   a. T Tauri stars
   b. Cepheids
   c. quasars
   d. planetary nebulae
   e. pulsars.

7. The core of a star remaining after the supernova event may become a
   a. quasar
   b. white dwarf
   c. neutron star
   d. black dwarf
   e. planet.

8. A gravitational red shift is
   a. caused by objects moving away from us
   b. caused by strong magnetic fields
   c. caused by the effect of a strong gravitational field on light
   d. observed as a star collapses to become a black hole.

9. In order for us to detect a pulsar
   a. a beam of particles must sweep across the earth's orbit
   b. it must be in a binary system
   c. the spectral lines must show Zeeman splitting
   d. all of the above.

10. The fate of a white dwarf is
    a. to eventually collapse to become a neutron star
    b. to eject a shell of material and become a planetary nebula
    c. to explode as a nova
    d. to cool until it no longer emits any light.

Answers to Selected Problems in the Text

2. Nova: Assume $10^{-4}M_\odot$ undergoes nuclear reactions in which 10% of the mass is converted to energy. Then the total mass converted to energy is $10^{-5}M_\odot$, or 2×10^28 grams ($1M_\odot = 2 \times 10^{33}$ g). The amount of energy released is:

$$E = mc^2 = (2 \times 10^{28}\ \mathrm{g})(3 \times 10^{10}\ \mathrm{cms}^{-1})^2 = 1.8 \times 10^{49}\ \mathrm{erg}$$

   Type I supernova: Assume $1M_\odot$ undergoes nuclear reactions, with the same 10% efficiency assumed for the nova. Since 10^4 times as much material is involved in reactions for the supernova as for the nova, the energy released is 10,000 times greater, or about 10^{53} ergs.

4. The wavelength of maximum emission from a 10^6 K pulsar is found from Wein's law:

$$\lambda_{max} = \frac{3 \times 10^7}{T}\ \text{Å} = \frac{3 \times 10^7}{10^6} = 30\ \text{Å}$$

   The dominant emission is in the X-ray portion of the spectrum.

6. We find the Schwarzschild radii for the objects simply by plugging into the formula:

$$R_s = \frac{2GM}{c^2}$$

   where $G = 6.67 \times 10^{-8}$ in cgs units.
   (a) The Sun: Plugging in the mass of the Sun (2×10^{33} grams), we find a Schwarzschild radius of 3×10^5 cm, or 3 km.
   (b) A $20M_\odot$ star: the Schwarzschild radius is directly proportional to the mass. Thus, a twenty solar mass star has a Schwarzschild radius twenty times larger than that of the Sun, or 60 km.
   (c) Milky Way galaxy: The given mass of approximately (10^{45} grams) is about 5×10^{11} solar masses. The Schwarzschild radius of the galaxy therefore is about 5×10^{11} times larger than that of the Sun, or about 1.5×10^{12} km (about .16 light-year).

Answers to Self-Test

1–d, 2–c, 3–d, 4–a, 5–b, 6–e, 7–c, 8–c, 9–a, 10–d

Additional Readings

Anderson, L. 1976, X-rays from degenerate stars, *Mercury*, **5**(5):2.

Bell Burnell, S. J. 1978, Little green men, white dwarfs, or what?, *Sky and Telescope*, **55**(3):218.

Black, D. L. 1976, Black holes and their astrophysical implications, *Sky and Telescope*, **50**(1):20 (Part I); **50**(2):87 (Part II).

Clark, G. W. 1977, X-ray stars in globular clusters, *Scientific American*, **237** (4): 42.

Gursky, H. and van den Heuvel, E. P. J. 1975, X-ray emitting double stars, *Scientific American*, **232**(3):24.

Helfand, D. J. 1977, Pulsars: physics laboratories in our galaxy, *Mercury*, **6**(3):2.

Helfand, D. J. 1978, Recent observations of pulsars, *American Scientist*, **66**:332.

Irwin, J. 1978, The case of the degenerate dwarf, *Mercury*, **7**(6):125.

Kaufmann, W. 1974, Black hole, white holes, and worm holes, *Mercury*, **3**(3):26.

Lewin, W. H. G. 1977, X-ray outbursts in our galaxy, *American Scientist*, **65**:605.

Lewin, W. H. G. 1981, The sources of celestial X-ray bursts, *Scientific American*, **244**(5):72.

Lewin, W. H. G. and van Paradijs, J. 1979, What are X-ray bursters?, *Sky and Telescope*, **57**(5):446.

Margon, B. 1979, SS 433: one of a kind?, *Mercury*, **8**(5):108.

Margon, B. 1980, The bizarre spectrum of SS433, *Scientific American*, **243**(4):54.

Pacini, F. and Rees, M. J. 1973, Rotation in high-energy astrophysics, *Scientific American*, **228**(2):98.

Penrose, R. 1972, Black holes, *Scientific American*, **226**(5):38.

Peters, P. C. 1974, Black holes: new horizons in gravitational physics, *American Scientist*, **62**:575.

Ruderman, M. A. 1971, Solid stars, *Scientific American*, **224**(2):24.

Ruffini, R. and Wheeler, J. A. 1971, Introducing the black hole, *Physics Today*, **24**(1):30.

Sexl, R. and Sexl, H. 1980, Curved space-time near a neutron star, *Mercury*, **9**(2):38.

Smarr, L. L. and Press, W. H. 1978, Spacetime: black holes and gravitational waves, *American Scientist*, **66**:72.

Stokes, G. M. and Michalsky, J. J. 1979, Cygnus X-1, *Mercury*, **8**(3):60.

Strong, I. B. and Klebesadel, R. W. 1976, Cosmic gamma-ray bursts, *Scientific American*, **235**(4):66.

Thorne, K. S. 1974, The search for black holes, *Scientific American*, **231**(6):32.

Van Horn, H. M. 1979, The physics of white dwarfs, *Physics Today*, **32**(1):23.

Wheeler, J. C. 1973, After the supernova, what?, *American Scientist*, **61**:42.

Wheeler, J. C. 1975, Black holes, neutron stars, and the green, green grass, *Griffith Observer*, **39**(5):2.

Chapter 25

Structure and Organization of the Milky Way

Untangling the structure of our Milky Way Galaxy has not been easy, because we are inside it and have no way to see the galaxy as a whole. In addition, problems such as interstellar extinction caused by gas and dust have made many observations more complicated. Despite these obstacles, we have made great progress in the last 50 years.

Our galaxy is pancake-shaped with a central bulge. The sun is located about two-thirds of the way out from the center. From the outside, the most distinguishing characteristic of the galaxy would be the spiral arms winding out through the disk. They define the location of the brightest stars, H II regions and young stars. There are, however, many dimmer stars between the arms and in the halo of the galaxy. The central bulge and inner part of the galaxy rotate like a solid body, while the outer part rotates differentially, following Kepler's third law. Stellar motions lead us to a determination of the minimum mass of the galaxy of roughly 10^{11} solar masses.

Because the galaxy is so huge, standard distance-determination techniques (such as trigonometric parallax) fail, and new methods are needed. Cepheid variables and RR Lyrae variables, which exhibit a period-luminosity relation, have been used to determine the size scale of the galaxy. These luminous stars have pulsational periods that are related to their absolute magnitudes.

The size of the galaxy and the sun's location within it were eventually determined from observations of RR Lyrae variables in globular clusters, which revealed that these clusters are distributed about a point far from the sun; and from the analysis of stellar motions near the sun, which showed that stars in our vicinity are orbiting a distant galactic center.

Even the brightest stars are inadequate for tracing spiral arms, but the discovery of 21-cm radiation from hydrogen made it possible to map gas clouds to distances of tens of kiloparsecs. Since this gas tends to concentrate in the arms, we emerge with a picture of spiral structure.

In recent years new aspects of our galaxy have been revealed, including an active core and a very extended halo, containing as much as 90 percent of the total mass. Various observations indicate that there may be a central object with a mass of 10^6 solar masses, whose nature is uncertain.

Words and Phrases

Cepheid variables	nucleus of the galaxy
differential rotation	peculiar motion
halo of the galaxy	period-luminosity relation
interstellar extinction	solar motion
kiloparsec	star counts
Milky Way	21-cm radiation

Exercise: Galactic Structure

Although we cannot travel outside our galaxy to look down on the disk, we can construct a map or picture from the location and distance of different stars. We measure positions in the galaxy by specifying the angular distance of objects from fixed reference directions, just as we describe the positions of stars by using right ascension and declination. This time, our reference frame is the plane of the galaxy. All positions are measured in degrees around the plane, starting at the center. Thus 0° is towards the galactic center, 90° is in the direction of rotation, 180° is towards the anticenter, and 270° is away from the direction of rotation. We also need to give the angular position north or south of the galactic plane, in order to completely specify a star's position in galactic coordinates. Using galactic coordinates and the distance of various stellar groups, we will construct a picture of spiral arms in the vicinity of the sun. As we will be working only with objects in the galactic plane, the angular measure north or south of the plane will always be zero and can be ignored.

The following table gives the angular distance from the galactic center, called the galactic longitude, and the linear distance from the sun for a number of OB associations. Plot these, using either regular graph paper or polar coordinate graph paper if it is available, with the sun at the center. Answer the following questions about your figure:

a. On your plot, sketch the outline of the spiral arms
b. Where is the center of the galaxy in your figure? It will probably be off the paper.
c. How do you suppose the distance of each of these OB associations was found? How many separate observations were necessary if each association has an average of 15 stars?
d. Why do you suppose that there are no associations more distant than 4.4 kpc in this list?

Galactic OB Associations*

Galactic Longitude	Distance (kpc)	Galactic Longitude	Distance (kpc)
10°	1.6	158°	0.4
11	1.7	173	1.3
12	2.4	173	3.2
16	2.2	188	1.5
17	1.7	205	0.7
18	2.0	206	0.5
23	1.0	207	1.5
60	2.0	224	1.3
72	2.3	238	1.5
75	1.8	244	2.5
78	2.3	246	4.4
78	1.2	266	1.8
80	1.8	286	2.5
88	0.8	287	3.5
97	0.6	287	3.6
100	0.8	287	2.6
103	3.5	287	2.5
108	2.1	290	2.0
110	0.9	291	3.2
112	2.6	294	2.4
116	2.5	303	2.5
118	0.8	320	4.0
120	2.9	332	4.0
122	2.5	333	1.6
129	2.9	339	2.5
134	2.3	340	1.4
136	2.2	340	3.5
143	1.0	343	1.9
146	3.3	351	2.6

* Data from Humphreys, R. 1978, *Astrophysical Journal Supplements*, **38**:309.

Self Test

1. The metal content of globular cluster stars is much lower than that of solar-type stars because
   a. these stars cannot produce elements heavier than carbon
   b. globular cluster stars were formed far from the plane of the galaxy
   c. globular cluster stars were formed so long ago that no metal was available
   d. all of the above.

2. The sun is presently on the edge of a spiral arm of the galaxy. It will:
   a. remain in the same place relative to the spiral arm for the rest of its career
   b. orbit the galaxy and repeatedly pass through spiral arms
   c. move in towards the center of the galaxy along the spiral arm
   d. none of the above.

3. Which statement about globular clusters is true?
   a. they were all formed in the center of our galaxy
   b. they contain stars with high metal abundance
   c. they are associated with gas and dust
   d. they are found in the galactic halo.

4. The sun's galactic motion is toward the center of the galaxy
   a. correct
   b. wrong: the sun is moving away from the galactic center
   c. wrong: the sun's motion is an elongated ellipse
   d. wrong: the sun moves in a circular orbit.

5. The time it takes the sun to travel around the galaxy is about _____years
   a. 230
   b. 2×10^5 years
   c. 2×10^8 years
   d. 2×10^{11} years.

6. According to modern observations, what is the approximate distance to the center of the galaxy?
   a. 93 million km
   b. 10 parsecs
   c. 10 kiloparsecs
   d. 30 kiloparsecs.

7. In a distant group of stars there is a Cepheid variable. What observations do you need to make to find the distance to the group of stars?
   a. apparent magnitude only
   b. period only
   c. a and b only
   d. a and b and chemical composition
   e. d and the radius.

8. The distribution of B stars in the galactic plane shows a spiral arm pattern because:
   a. such luminous objects do not live long enough to move out of the arms
   b. their great masses do not let them overcome the attraction of the arms
   c. they continue to accumulate new material from the clouds in the arms
   d. none of the above: the distribution of stars does not show a spiral pattern.

9. Radio astronomers have mapped the distribution of hydrogen in the plane of the galaxy by
   a. measuring the distances to the clouds by parallax
   b. relating radial velocities to distance
   c. measuring parallaxes of stars behind the clouds
   d. all of the above.

10. In determining the shape of our galaxy from the distribution of globular clusters, we assume:
   a. the distance to all globular clusters is the same
   b. the globular clusters are distributed uniformly around the galactic center
   c. that globular clusters outline the spiral arms of the galaxy
   d. that globular clusters were all ejected from the center of the galaxy at about the same time.

Answers to Selected Problems from the Text

2. The Cepheid has an apparent magnitude of +16, and an absolute magnitude of −4 (from the period-luminosity relation). Its distance is:

$$d = 10^{1+\frac{m-M}{5}} \text{ pc} = 10^{1+\frac{16-(-4)}{5}} \text{ pc} = 10^{5} \text{ pc}$$

Clearly, the use of Cepheid variables as distance indicators allows us to extend the distance measurements far beyond that possible with parallax.

4. We calculate the orbital period (P) of the Sun about the galaxy by assuming a circular orbit with a radius of 10,000 pc and a speed of 250 $km\ s^{-1}$:

$$\mathrm{P} = \frac{\text{orbit circumference}}{\text{speed}} = \frac{2\pi(10^4\ \mathrm{pc})(3.26\ \mathrm{pc/ly})(9.46\times 10^{12}\ \mathrm{km/ly})}{250\ \mathrm{km/s}}$$

$$\Longrightarrow P = 7.8\times 10^{15}\ \mathrm{s} = 250\ \text{million years}$$

6. We follow the method of problem 4 to find the orbital period of material at a distance of 5 kpc from the galactic center with an orbital speed of 150 $km\ s^{-1}$:

$$\mathrm{P} = \frac{2\pi(5\times 10^3\ \mathrm{pc})(3.26\ \mathrm{pc/ly})(9.46\times 10^{12}\ \mathrm{km/ly})}{150\ \mathrm{km/s}}$$

$$\Longrightarrow P = 6.5\times 10^{15}\ \mathrm{s} \approx 200\ \text{million years}$$

We can calculate the mass of the galaxy interior to 5 kpc in the galaxy with Kepler's third law:

$$\mathrm{p}^2 = \frac{4\pi^2}{\mathrm{G}(\mathrm{M}_1+\mathrm{M}_2)}\mathrm{a}^3$$

The two masses are the mass of the galaxy (interior to 5 kpc) and the mass of an orbiting object; the latter may be neglected since it is negligible compared to the former. We plug in the period in seconds, the distance in centimeters (5000 pc $\approx 1.5\times 10^{22}$ cm), and $\mathrm{G} = 6.67\times 10^{-8}$ (cgs units): the mass is about 6×10^{43} grams, or about 30 billion solar masses.

8. We are asked to choose the first five stars in the list of the brightest stars (Appendix 8), and determine what their apparent magnitudes would be if they were located at a distance of only 0.1 pc (the average distance between stars in a globular cluster). Appendix 8 gives the actual apparent magnitudes of the stars, and their distances from Earth. To solve this problem, we first must determine the absolute magnitudes of the stars, and then find their apparent magnitudes at a distance of 0.1 pc.
Let m_0 be the actual apparent magnitude of a star (in our sky), and m be the apparent magnitude if the star was located only 0.1 pc away. Let d_0 be the actual distance of the star from Earth, and M is the absolute magnitude. Using

the relation between apparent magnitude, absolute magnitude, and distance (see solution to problem 1, chapter 20), we have:

$$M = m_0 - 5\log\frac{d_0}{10\ \text{pc}}$$

$$\text{and, } m = M + 5\log\frac{0.1\ \text{pc}}{10\ \text{pc}} = M - 10$$

We can combine the above relations to solve for the apparent magnitude at a distance of 0.1 pc:

$$m = m_0 - 5\log\frac{d_0}{10\ \text{pc}} - 10$$

We now find the apparent magnitudes for the first five stars in the list:

α Eri (Archernar; $m_0 = 0.51$, $d_0 = 36$ pc):

$$m = 0.51 - 5\log\frac{36\ \text{pc}}{10\ \text{pc}} - 10 = -12.3$$

α UMi (Polaris; $m_0 = 1.99$, $d_0 = 208$ pc):

$$m = 1.99 - 5\log\frac{208\ \text{pc}}{10\ \text{pc}} - 10 = -14.6$$

α Per (Mirfak; $m_0 = 1.80$, $d_0 = 175$ pc):

$$m = 1.80 - 5\log\frac{175\ \text{pc}}{10\ \text{pc}} - 10 = -14.4$$

α Tau (Aldebaran; $m_0 = 0.86$, $d_0 = 21$ pc):

$$m = 0.86 - 5\log\frac{21\ \text{pc}}{10\ \text{pc}} - 10 = -10.8$$

β Ori (Rigel; $m_0 = 0.14$, $d_0 = 276$ pc):

$$m = 0.14 - 5\log\frac{276\ \text{pc}}{10\ \text{pc}} - 10 = -17.1$$

Except for Aldebaran, all of the stars above would be brighter than the full moon ($m \approx 12$) if located 0.1 pc away; Rigel would be 100 times (5 magnitudes) brighter than the full Moon. Of course, no blue supergiants (like Rigel) would be found in a globular cluster; many luminous red giants, however, would be expected. The night sky on a planet in a globular cluster would be quite bright.

Answers to Self Test

1–c, 2–b, 3–d, 4–d, 5–c, 6–c, 7–c, 8–a, 9–b, 10–b

Additional Readings

Bok, B. J. 1972, Updating galactic spiral structure, *American Scientist*, **60**:708.

Bok, B. J. 1981, The Milky Way galaxy, *Scientific American*, **244**(3):92.

Geballe, T. R. 1979, The central parsec of the galaxy, *Scientific American*, **241**(1):52.

Herbst, W. 1982, The local system of stars, *Sky and Telescope*, **63**(6):574.

Iben, I. 1970, Globular-cluster stars, *Scientific American*, **223**(1):26.

Percy, J. R. 1975, Pulsating stars, *Scientific American*, **232**(6):66.

Sanders, R. H. and Wrixon, G. T. 1974, The center of the galaxy, *Scientific American*, **230**(4):66.

Wynn-Williams, G. 1979, The center of our galaxy, *Mercury*, **8**(5):97.

Chapter 26

The Interstellar Medium

It is a testimony to the vast distances encountered in space that a medium of gas and dust that is so thin it is superior to the best man-made vacuum can severely limit our view of the galaxy we live in. The interstellar medium, 99% gas and 1% dust (by mass), is also the reservoir for star formation, and its chemical composition reflects the processes in which stars convert light elements to heavier ones.

The existence of interstellar dust has been recognized for a long time. Photographs show dark areas of dense clouds, and light from distant stars is increasingly absorbed with decreasing wavelength. The result is to make the star's color redder. The color excess, a measure of this interstellar reddening, is an important quantity to know for most stellar observations. From the way light is reddened, we deduce the size of the grains, and from the polarization of starlight we conclude that the grains are not all spherical, but elongated. We do not know very much about the composition or origin of the grains, however. The most reasonable supposition is that they are silicates or graphite which have formed around newborn stars or cool supergiant stars in the process of shedding their outer layers.

Interstellar gas manifests itself in many ways: visible clouds heated by a nearby star until they glow, 21-cm radiation from diffuse clouds detected by radio telescopes, and absorption lines seen in both ultraviolet and visible wavelengths. The composition of the gas is similar to the sun and other stars. It is largely hydrogen, but has traces of other elements. Dark clouds, so dense that no starlight can be seen through them, are important sites for star formation. Molecules produce line radiation much as atoms do, although the lines are at much longer wavelengths. Radio techniques have detected over 70 different molecules within these clouds, including organic molecules, similar to those found in living material.

The interstellar medium is not a tranquil place. Clouds, often associated with supernova remnants, are found moving with speeds up to 100 km/sec. X-ray and ultraviolet observations reveal gas as hot as a million degrees, although the density is exceedingly low. What is the source of this agitation and heating? Primarily it is supernovae, which occur every decade or so in the galaxy. Another contributor is the stellar winds from hot stars, which can sweep up material and evacuate large interstellar cavities surrounding such stars.

Words and Phrases

color excess
diffuse cloud
emission nebula
H II region
interstellar absorption line
interstellar cloud
interstellar dust grain
interstellar extinction
organic molecule
polarization of star light
reflection nebula

Exercise: Interstellar Clouds

The 21-cm radiation from diffuse clouds is a powerful tool for mapping the galaxy. Since interstellar absorption is minimal in the radio region, these clouds can be detected at great distances. The doppler shift of the radiation is a measure of a cloud's velocity with respect to us. The determination of a cloud's distance cannot be made directly, however. Rather, it depends on our understanding of the overall rotation of the galaxy. From this understanding, or model of galactic rotation, we can predict the velocity as a function of distance from the center of the galaxy. We then find the distance of a cloud from its measured Doppler shift.

The actual analysis requires a bit more computation than is appropriate here, but we can do a simple approximation to it. In the exercise in Chapter 25 we used galactic coordinates to locate objects with respect to the galactic center. Now suppose that we have the galactic coordinates of clouds observed to have 21-cm radiation, and the Doppler shifts of these clouds, as listed on the next page. Assuming Keplerian motion, sketch the galactic plane and estimate the position of each cloud. To do this, you must use Kepler's third law to calculate the orbital period, and from that the orbital velocity, at various distances from the galactic center. Then, by looking at your drawing, you can estimate what the radial (i.e. line-of-sight) velocity would be, as seen from the earth, for clouds at various positions along the directions specified in the table. If the rotation of the galaxy were solid body rather than Keplerian, how would you expect these Doppler shifts to appear?

Diffuse Clouds

Cloud Number	Galactic Longitude	Radial Velocity (km/sec)
1	0°	0
2	30	+10
3	35	+20
4	110	-10
5	115	-20
6	135	-10
7	200	+10
8	210	+20
9	290	-10
10	300	-20

Self-Test

1. From the polarization of starlight we learn
   a. the amount of interstellar dust between us and a star
   b. the chemical composition of the interstellar medium
   c. about the magnetic field of the galaxy
   d. the average size of the interstellar grains.

2. Approximately what percent of the mass of the galaxy is contained in the interstellar medium?
   a. 10%
   b. 5%
   c. 1%
   d. 0.1%.

3. The most complex type of molecule found so far in space is
   a. an amino acid
   b. an organic molecule
   c. molecular hydrogen
   d. a bacterium.

4. The 21-cm emission line that radio telescopes can detect is formed
   a. when electrons in hydrogen gas drop from one orbit to a lower one
   b. when a spinning electron spontaneously reverses its direction of spin
   c. when electrons spiral around the lines of force in a magnetic field
   d. none of the above.

5. Stars that are far away generally appear
   a. redder
   b. bluer

   than closer stars. This is because
   a. red stars are intrinsically brighter than blue stars
   b. the spectra of the distant stars are redshifted
   c. there is gas and dust between the stars which absorbs blue light.

6. Where have radio astronomers found complex molecules in space?
   a. in H II regions
   b. around supernova remnants
   c. in dense interstellar dust clouds
   d. all of the above.

7. The amount of reddening of starlight by dust can be measured because dust does not affect the star's
   a. apparent magnitude
   b. color
   c. absolute magnitude
   d. spectral lines
   e. blue wavelengths.

8. The type of interaction between interstellar matter and radiation depends primarily on
   a. composition of particles
   b. sizes of particles relative to radiation wavelength
   c. source of the radiation
   d. both a and b
   e. all three, a, b, and c.

9. In the interstellar medium, the average particle density is close to
   a. 0 particles/cm^3
   b. 1 particle/cm^3
   c. 100 particles/cm^3
   d. $10^4/cm^3$
   e. $10^{15}/cm^3$.

10. What effect does interstellar dust have on the magnitudes and colors of a star?
    a. dims the star
    b. makes the star appear redder
    c. makes the star dimmer and redder
    d. no change.

Answers to Selected Problems from the Text

1. Interstellar dust is causing a particular star to appear only 1/100 as bright as it would otherwise. Since a factor of 100 in brightness corresponds to a change of 5 magnitudes, we say that the dust is causing interstellar extinction of 5 magnitudes.

3. In this problem we estimate the average density of the interstellar medium by dividing the interstellar mass by the volume of the galactic disk. We are told to assume that the galactic disk has a radius (r) of 15,000 pc; a thickness (z)

of 200 pc; and a mass of $2 \times 10^{11} M_\odot$, 15% of which is interstellar matter. We assume that the disk has a cylindrical geometry, so that its volume (V) is equal to its area (πr^2) times its thickness. Before calculating the density, we first convert to units of grams and cm:

$$M_{ISM} = .15 \times (2 \times 10^{11} M_\odot) \times (2 \times 10^{33}\ \text{g}/M_\odot) = 6 \times 10^{43}\ \text{g}$$

$$r_{disk} = (15,000\ \text{pc})(3.26\ \text{pc/ly})(9.46 \times 10^{17}\ \text{cm/ly}) = 4.6 \times 10^{22}\ \text{cm}$$

$$z_{disk} = (200\ \text{pc})(3.26\ \text{pc/ly})(9.46 \times 10^{17}\ \text{cm/ly}) = 6.2 \times 10^{20}\ \text{cm}$$

The density (ρ) of interstellar material, then, is:

$$\rho = \frac{M_{ISM}}{V_{disk}} = \frac{6 \times 10^{43}\ \text{g}}{\pi(4.6 \times 10^{22}\ \text{cm})^2(6.2 \times 10^{20}\ \text{cm})}$$

$$\Longrightarrow \rho = 1.5 \times 10^{-23}\ \text{g cm}^{-3}$$

Most of the interstellar material is, of course, in the form of hydrogen. We recall that there are about 6×10^{23} (Avogadro's number) hydrogen atoms in 1 gram of hydrogen gas. Thus, if the interstellar material were entirely hydrogen, there would be about 9 atoms per cubic centimeter:

$$(1.5 \times 10^{-23}\ \text{g cm}^{-3}) \times (6 \times 10^{23}\ \text{atoms g}^{-1} = 9\ \text{atoms cm}^{-3}$$

The actual number density would be somewhat smaller, since about one-fourth of the material is in the form of helium (and a much smaller fraction in heavier elements or dust grains).

Answers to Self Test

1–d, 2–c, 3–c, 4–b, 5–a–c, 6–c, 7–a, 8–b, 9–b, 10–c

Additional Readings

Blitz, L. 1982, Giant molecular cloud complexes in the galaxy, *Scientific American*, **246**(4): 84.

Chaisson, E. J. 1978, Gaseous nebulas, *Scientific American*, **239**(6): 164.

Charles, P. A. and Culhane, J. L. 1975, X-rays from supernova remnants, *Scientific American*, **233**(6): 38.

Chevalier, R. A. 1978, Supernova remnants, *American Scientist*, **66**: 712.

Cowsik, R. and Price, P. B. 1971, Origins of cosmic rays, *Physics Today*, **24**(10): 30.

Dickenson, D. F. 1978, Cosmic masers, *Scientific American*, **238**(6): 90.

Gordon, M. A. and Burton, W. B. 1979, Carbon monoxide in the galaxy, *Scientific American*, **240**(5): 54.

Herbig, G. H. 1974, Interstellar smog, *American Scientist*, **62**: 200.

Heiles, C. 1978, The structure of the interstellar medium, *Scientific American*, **238**(1): 74.

Herbst, E. and Klemperer, W. 1976, The formation of interstellar molecules, *Physics Today*, **29**(6): 32.

Jura, M. 1977, Interstellar clouds and molecular hydrogen, *American Scientist*, **65**: 446.

Kaler, J. B. 1982, Bubbles from dying stars, *Sky and Telescope*, **63**(2): 129.

Knacke, R. F. 1979, Solid particles in space, *Sky and Telescope*, **57**(4): 347.

Linsley, J. 1978, The highest-energy cosmic rays, *Scientific American*, **239**(1): 60.

Maran, S. P. 1971, The Gum nebula, *Scientific American*, **225**(6): 20.

Meyer, P. Ramaty, R. and Webber, W. R. 1974, Cosmic-ray astronomy with energetic particles, *Physics Today*, **23**(10): 23.

Miller, J. S. 1974, The structure of emission nebulas, *Scientific American*, **231**(4): 34.

Snow, T. P. 1976, The interstellar medium: much ado about nothing, *Griffith Observer*, **40**(9): 2.

Solomon, P. M. 1973, Interstellar molecules, *Physics Today*, **26**(3): 32.

Terzian, Y. 1977, Recent findings about planetary nebulae, *Sky and Telescope*, **54**(6): 459.

Turner, B. R. 1973, Interstellar molecules, *Scientific American*, **228**(3): 50.

Chapter 27

The Formation and Evolution of the Galaxy

The life history of stars is linked intimately to the evolution of the galaxy to which they belong. We can understand the present state of our galaxy only with our knowledge of stellar evolution, and vice-versa. Our galaxy consists of a disk and a central bulge. Through the disk wind spiral arms containing bright, young, massive O and B stars and their associated H II regions. There are a number of differences between the stars in the disk and the central bulge. Those in the bulge and the galactic halo have relatively low heavy element abundances. Stars in the vicinity of the sun have high abundances of heavy elements: iron, for example, is about 10^{-5} as abundant as hydrogen in the sun, but 100 times less abundant than this in the halo stars. These differences have been used to divide stars into two classes: Population II for stars in the halo and nucleus, and Population I for those in the spiral arms. The sun is typical of Pop. I. Further study have shown that intermediate population types exist, so the division may be akin to classifying people as old or young, ignoring all those in between, but the concept of population type is nevertheless useful.

There are some Pop. II stars in the solar vicinity, distinguished not only by their low metal content, but also by highly elliptical orbits which carry them far from the plane of the galaxy. A scenario to explain these observations emerges when we recall that material is cycled back and forth between stars and the interstellar medium, with increasing heavy element enrichment in each new generation. Stars formed early in the history of the galaxy have low abundances of heavy elements. Therefore Pop. II stars are old. There are also abundance variations with location, and the decreasing heavy element abundance with distance from the galactic center reflects a decreasing rate of star formation.

Spiral arms are the most distinctive features of galaxies like our own, and also the hardest to explain. Clearly they are not simply trailing streamers, or they would have wound up tightly long ago. The spiral density wave theory of 1960 was the first successful attempt to explain them. This theory assumes that the large-scale organization of the galaxy is imposed by a wave motion akin to sound waves. In this theory, a spiral-shaped wave pattern centered on the galactic nucleus creates a pinwheel of alternating dense and rarified regions. The waves affect the interstellar medium the most, so gas and dust are concentrated in the dense regions of the waves. This leads to star formation in these regions. The most luminous stars, O and B types, are so short lived that they do not last long enough to leave the density wave in which they were born. Hence the density condensations, or spiral arms, are distinguished by young luminous stars as well as gas and dust.

Our current evidence suggests that the Milky Way originated as a gigantic gas cloud. This cloud began to collapse as star formation started. The first stars to form (Pop. II) had randomly oriented orbits, but as the cloud collapsed to a disk-shaped configuration, later stars (Pop. I) were born with orbits in this disk. Probably the spiral arms formed at this stage, although the mechanism for setting up a density wave is very uncertain.

Words and Phrases

abundance gradients
density wave theory
Populations I and II
heavy elements
high velocity stars
disk population

Exercise: Galactic Evolution

The galaxy is about 15 billion years old. Its appearance has changed greatly over this interval. Write an essay describing in detail the appearance of the galaxy, seen from a distance of 10 kpc from its center, at the following times: 15 billion years ago, 10 billion years ago, and the present time.

Self Test

1. The age of our galaxy is about
   a. 15×10^6 years
   b. 15×10^8 years
   c. 15×10^9 years
   d. 15×10^{10} years.

2. The age of our galaxy is found
   a. from the velocity of the sun about the galactic center and the number of trips it has completed
   b. from radioactive dating
   c. from the maximum ages of globular cluster stars
   d. from mathematical models of stellar evolution.

3. The sun has made about _____ orbits of the Milky Way
   a. two
   b. five
   c. fifteen
   d. fifty.

4. The masses of Pop. II stars are all less than about 1 solar mass. This is because
   a. formation of massive stars requires a higher concentration of heavy elements than were available where Pop. II stars formed
   b. massive Pop. II stars have long since ended their lives
   c. mass loss through a stellar wind has reduced the massive stars to less than one solar mass
   d. wrong: Pop. II stars include significant numbers of massive stars.

5. The differences between Population I and II are
   a. Pop. I stars have low metal abundance and highly elliptical orbits
   b. Pop. I has low metal abundance and Pop. II has high metal abundance
   c. Pop. II stars have low metal abundance and highly elliptical orbits
   d. Pop. I stars are older than Pop. II stars.

6. An analogy in the solar system to the contrast between Pop. I and II stars is that
   a. Pop. I corresponds to comets, Pop. II to planets
   b. Pop. I corresponds to planets, Pop. II to comets
   c. Pop. I corresponds to planets, Pop. II to asteroids
   d. Pop. I corresponds to comets, Pop. II to asteroids.

7. In the density wave theory of spiral structure,
   a. stars move more slowly through the spiral arms than through the inter-arm regions
   b. stars move outward, along the spiral arms
   c. the relative position of stars and spiral arms remain fixed
   d. stars form in the inter-arm regions.

8. The largest unsolved problem about our galaxy is
   a. how it reached its present flattened shape
   b. why globular clusters exist far from the galactic plane
   c. the distance to the galactic center
   d. why spiral arms exist.

9. When the galaxy was young
   a. Pop. II stars were forming
   b. the sun had only recently formed
   c. it contained fewer heavy elements
   d. it had the characteristics of a large elliptical galaxy
   e. it contained a larger amount of gas than now.

10. The spiral arms of our galaxy
   a. are composed of stars; very few stars are found between the arms
   b. are the location of the brightest and youngest stars
   c. are gradually winding up as the galaxy evolves into an elliptical galaxy
   d. all of the above.

Answers to Selected Problems from the Text

1. The spectrum of a particular star reveals the presence of iron in the ratio of 1:40,000 (by number) compared to hydrogen. As a fraction, this ratio is 2×10^{-5}, an iron abundance nearly identical to that of the Sun. Thus, the star is a member of population I.

3. The problem asks how many times the Sun has passed through a spiral arm in the 15 billion year lifetime of the galaxy, assuming that it passes through a spiral arm every 5×10^8 years. A simple division shows that, in 15 billion years, there would be 30 passes through a spiral arm. Of course, the Sun has not existed since the beginning of the galaxy; it only is about 5 billion years

old. Therefore, the Sun could have passed through a spiral arm only about 10 times.

Answers to Self Test

1–c, 2–c, 3–d, 4–b, 5–c, 6–b, 7–a, 8–d, 9–a–c–e, 10–b

Additional Readings

Bok, B. J. 1981, Our bigger and better galaxy, *Mercury*, **10**(5):130.

Larson, R. B. 1979, The formation of galaxies, *Mercury*, **8**(3):53.

Shu, F. H. 1973, Spiral structure, dust clouds, and star formation, *American Scientist*, **61**:524.

Weaver, H. 1975/1976, Steps towards understanding the large-scale structure of the Milky Way, *Mercury*, **4**(5):18 (Part I); **4**(6):18 (Part II); **5**(1):19 (Part III).

Chapter 28

The Nature of the Nebulae

In the constellation of Andromeda there lies a faint, fuzzy patch that can be seen on a clear moonless night. Sometimes called the Andromeda nebula, it typifies the many such objects (all the others are fainter) noted by astronomers early in this century. For a long time, the nature of these nebulae was unclear, and they were believed to be gas clouds within our own galaxy. It was not until 1924 that it was realized that they were separate galaxies like our own Milky Way galaxy. The confusion between gas clouds and separate galaxies or "island universes" arose principally because the galaxies are so distant that individual stars could not easily be resolved in them. The controversy was settled when Cepheid variables were identified in a few of the nebulae, and from the period-luminosity relation, Edwin Hubble demonstrated that these stars were in distant galaxies.

Hubble classified the galaxies systematically according to their shape. The elliptical galaxies were subdivided according to their degree of flattening. Spiral galaxies were divided into those showing a central bar and those without one, and further classified according to the tightness with which the arms wound around the nucleus. For many years it was thought that these types represented an evolutionary sequence, but eventually it became clear that both spiral and elliptical galaxies contain old stellar populations, and that the differences in galactic type could not be explained as an age effect.

The first problem in understanding galaxies is to find their distances. The methods developed for stars in our own galaxy will not work for galaxies, except for observations of Cepheid variables in a few nearby ones. In general, however, any luminous objects whose absolute magnitude is known can be used to estimate distances. These "standard candles" include supergiants, supernovae, H II regions and star clusters.

An application of Kepler's third law leads to the determination of the mass of a distant galaxy. This requires that the velocity of orbiting objects at some distance from the center be known, or that one observe a pair of orbiting galaxies. Galaxies exhibit a tremendous range in luminosity, up to a factor of a million, and diameters that vary by a factor of one hundred.

In general it seems that spiral galaxies are currently forming stars and cycling matter through the interstellar medium, while ellipticals are static. Apparently, the form a galaxy takes depends on the rate of star formation in its early history. If the gas in the original cloud from which the galaxy formed was rapidly converted to stars, an elliptical galaxy resulted. Otherwise the gas would have time to form a disk, and a spiral galaxy would be formed.

Words and Phrases

barred spirals
dwarf ellipticals
elliptical galaxy
Hubble
irregular galaxy
mass-to-light ratio
megaparsec
nebulae
spiral galaxies
standard candle
velocity curve
velocity dispersion

Exercise: Large Numbers

Astronomy is a science of large numbers, especially in discussions of extragalactic objects. Distances measured in megaparsecs are difficult to comprehend, but so are the numbers associated with the national debt, or a major corporate budget. Sometimes these numbers are easier to grasp if they are reduced to something more meaningful. The names given to large numbers are often confusing: a thousand is 10^3, a million is 10^6, a billion is 10^9 and a trillion is 10^{12}, at least in the United States. The term zillion has no mathematical meaning. In most other countries the term trillion applies to 10^{18}, which in the U.S. is a quintillion. The use of scientific notation avoids the confusion caused by these terms. Remember the rules for working in scientific notation.

Consider some of the following large numbers. When a fast-food chain claims that it has sold 50 billion hamburgers, what does this mean if we assume every person in the U.S. has bought equal numbers of these hamburgers? There are about 250 million people, so

$$\frac{50 \times 10^9 \text{ hamburgers}}{25 \times 10^7 \text{ people}} = 200 \text{ hamburgers per person.}$$

Try the following. The sun is 1.5×10^8 km from the earth. If we travelled at 1000 km/hr (on a commercial jet), how long would it take to reach the sun? How long does it take light, travelling at 3×10^5 km/sec, to reach us from the sun? The nearest star is about 2×10^5 times as far as the sun. How long would it take to reach it by jet? At the speed of light?

The large Magellanic Cloud, a satellite galaxy to our Milky Way, is 55 kpc in distance. Remember that a parsec is 3.26 light years. How many years ago did light reaching us now leave the Large Magellanic Cloud? The Andromeda galaxy is 570 kpc away. For how long has light been travelling to reach us? (Incidentally, this is the most distant object that you can see without a telescope.) Finally, consider a cluster of galaxies in the constellation Ursa Major, at a distance of 280 Mpc. How long ago did the light now reaching earth leave that cluster?

Self Test

1. Different galactic shapes are thought to be
   a. different stages in the life cycle of a galaxy
   b. determined mainly by the amount of mass originally in the proto-galaxy
   c. purely random phenomena
   d. determined partly by the initial angular momentum of the proto-galaxy
   e. none of the above.

2. Elliptical galaxies are older than spiral galaxies.
   a. true: star formation has finished and there is little gas and dust
   b. true: spiral arms have wound up, leading to the elliptical shape
   c. false: both contain very old stars, although ellipticals have more of them
   d. false: ellipticals have not yet developed spiral arms.

3. To measure the distance to the Andromeda galaxy, astronomers could best use
   a. Cepheids
   b. the velocity-distance relation
   c. the total absolute magnitude of the galaxy
   d. the magnitude of the brightest star in the galaxy.

4. We believe that little star formation takes place in elliptical galaxies because
   a. we rarely see any supernova explosions in these galaxies
   b. they are so small that all of their mass has already been used to form stars
   c. we do not observe much gas and dust in these galaxies
   d. all of the above.

5. Which of the following is not a method of determining distances to galaxies?
   a. Cepheid period-luminosity relation
   b. measurements of annual parallax motions of galaxies
   c. assuming that all supernovae reach the same absolute magnitude at maximum brightness
   d. assuming that the brightest galaxy in any cluster of galaxies has the same intrinsic brightness

6. Spiral galaxies contain
   a. very little interstellar gas and dust
   b. mostly old, red stars
   c. mostly young, blue stars
   d. young blue stars in the nucleus and old red stars in the spiral arms
   e. old red stars in the nucleus and young blue stars in the spiral arms.

7. It was once thought that galaxies originated as ellipticals, evolved to spirals, and ended as irregulars. The best argument against this theory is
   a. irregular galaxies have too much mass to be descended from spirals
   b. ellipticals do not contain dust and gas from which new stars can form
   c. spiral galaxies have the oldest stars
   d. none of the above.

8. The distance to a nearby galaxy could be about
   a. 8,000 parsecs
   b. 27,000 parsecs
   c. 1.5 million parsecs
   d. 2 billion parsecs.

9. The spectra of stars in distant galaxies appear to be almost identical to the spectra of stars in our own galaxy. Why is this important?
   a. we can measure doppler shifts in these spectra to determine motions of these galaxies
   b. the laws of physics governing the formation of these galaxies are similar to the laws of physics governing the formation of our own galaxy
   c. the relative abundances of chemical elements in these galaxies may be assumed to be fairly similar to the relative abundances measured in our galaxy
   d. all of the above.

Answers to Selected Problems From the Text

1. We calculate the apparent magnitude of a supergiant star with absolute magnitude M = −8, if such a star were located at a distance of 100,000 pc, or at a distance of 1,000,000 pc. We use the relation:

$$m = M + 5\log\frac{d}{10\ \text{pc}}$$

At a distance of 100,000 pc, the apparent magnitude of the star would be m = +12. At a distance of 1,000,000 pc, the apparent magnitude of the star would be m = +17.

4. The rotation curve of a particular spiral galaxy shows that the orbital velocity is 150 $km\ s^{-1}$ at a distance of 15 kpc from the center. We can use Kepler's third law to find the mass of the galaxy (interior to 15 kpc), but first must determine the orbital period:

$$p = \frac{\text{orbit circumference}}{\text{speed}} = \frac{2\pi(15,000\ \text{pc})(3.26\ \text{pc/ly})(9.46\times 10^{12}\ \text{km/ly})}{150\ \text{km/s}}$$

$$\Longrightarrow p = 1.9\times 10^{16}\ \text{s} = 6\times 10^{8}\ \text{yr}$$

We solve Kepler's law (in units of years and AU) to find:

$$\frac{M}{M_\odot} = \frac{a^3}{p^2} = \frac{[(15,000\ \text{pc})(206,265\ \text{AU/pc})]^3}{(6\times 10^8\ \text{yr})^2} = 8\times 10^{10}$$

The mass of the galaxy interior to 15 kpc is about 800 million solar masses.

Answers to Self-Test

1–d, 2–c, 3–a, 4–c, 5–b, 6–e, 7–b, 8–c, 9–d

Additional Readings

Abell, G. O. 1975, Beyond the Milky Way, *Astronomy*, **3**(1):6.

Bertola, F. 1981, What shape are elliptical galaxies?, *Sky and Telescope*, **61**(5):380.

de Boer, K. and Savage, B. D. 1982, The coronas of galaxies, *Scientific American*, **247**(2):54.

Hirshfeld, A. 1980, Inside dwarf galaxies, *Sky and Telescope*, **59**(4):287.

Hodge, P. W. 1981, The Andromeda galaxy, *Scientific American*, **244**(1):88.

Irwin, J. B. 1980, The case of the spinning spirals, *Griffith Observer*, **44**(2):2.

Seeley, D. and Berendzen, R. 1980, Astronomy's great debate, *Mercury*, **9**(4):67.

Strom, S. E. and Strom, K. M. 1979, The evolution of disk galaxies, *Scientific American*, **240**(4):56.

Talbot, R. J., Jensen, E. B., and Dufour, R. J. 1980, Anatomy of a spiral galaxy, *Sky and Telescope*, **60**(1):23.

Toomre, A. and Toomre, J. 1973, Violent tides between galaxies, *Scientific American*, **229**(6):38.

Chapter 29

Clusters and Superclusters: The Distribution of Galaxies

Our study of the distances and distribution of galaxies has shown that these giant objects tend to be found in clusters. The nature of these clusters can tell us a great deal about the early universe as well as how galaxies interact with one another.

The Milky Way is a major member of a small cluster of galaxies known as the Local Group. About 30 galaxies, most of them dwarf ellipticals, are arranged rather haphazardly in a region of space about 800 kpc across. The nearest neighboring cluster is about 1100 kpc away. The only objects in the Local Group that are visible to the unaided eye are the Large and Small Magellanic Clouds, irregular galaxies that are satellites of the Milky Way, and the Andromeda Galaxy. Stars in the Magellanic Clouds have a lower fraction of heavy elements than do stars in our galaxy, and therefore provide an important laboratory for many of our ideas about stellar evolution. Equally important, these galaxies allow us to calibrate methods of deriving distance scales for more distant galaxies.

Many clusters of galaxies contain hundreds or thousands of members, and frequent interactions between galaxies take place. The central regions of these clusters contain a very high fraction of elliptical or SO galaxies. This is probably the result of near-collisions between the members, which strip away gas, dust and halos.

The masses of clusters of galaxies are of interest in determining the mass of the universe, and there are two ways of measuring them. The masses of the individual galaxies can be estimated. Alternatively, the mass can be found from the velocity dispersion of the cluster. This method always gives a larger answer for a cluster mass than what is found by adding the masses of the individual galaxies, suggesting that mass may be hiding in galactic halos or between the galaxies. Recent observations of X-rays from clusters of galaxies indicate the presence of a gas between the galaxies that is as hot as a hundred million degrees. Since iron is present in this gas, it must be material that was processed in stellar interiors before being expelled into space.

Galaxy formation is similar to star formation in that it begins with the collapse of a cloud of gas. The trigger for collapse of interstellar clouds, leading to star formation, is interstellar turbulence, created by either a density wave or a supernova shock wave. It is not so clear what trigger would cause a pre-galactic cloud to collapse, since it is not certain that turbulence was present in the early universe.

Words and Phrases

Andromeda galaxy
clusters of galaxies
giant elliptical galaxies
intergalactic medium
Local Group
Magellanic Clouds
superclusters

Exercise: Colliding Galaxies

In discussing clusters of galaxies, we have considered the effects of close encounters between galaxies. There are observed cases of galaxies in the process of colliding, and galaxies which have collided in the past. If galaxies in a cluster can collide with each other, it might seem likely that stars within a galaxy such as our own could also undergo frequent collisions. To understand why this is not an important process, let us calculate how crowded a cluster of galaxies is compared with the density of stars in the solar neighborhood.

Near the sun, the density of stars is about 0.06 stars per cubic parsec. What fraction of a cubic parsec is therefore filled with stars? If we take the radius of the sun as typical, we can calculate its volume, equal to $\frac{4}{3}\pi R^3$, to estimate the volume of star material per cubic parsec. The radius of the sun in parsecs is about 2×10^{-8} pc, so the volume is about 3×10^{-23} pc^3. The fractional volume of a cubic parsec of space that is occupied by stars is therefore $0.06 \times (3 \times 10^{-23})$, or 2×10^{-24}. Compute the fractional volume occupied by stars in a globular cluster. Assume the cluster is a sphere 12 pc in diameter and contains 2×10^5 stars.

Now, compute the fractional volume occupied by galaxies in a rich cluster. Assume the cluster of galaxies is spherical and has a radius of 1 Mpc, or 1000 kpc. It contains 2500 galaxies and each galaxy is assumed spherical in shape, with a radius of 2 kpc.

Based on your calculations, which is more probable, collisions between galaxies in a cluster, or collisions between stars in a galaxy? What assumptions have we made that affect these conclusions?

Self Test

1. The percentage of ellipticals in rich clusters is _____the percentage in small clusters.
   a. equal to
   b. greater than
   c. less than
   d. not known.

2. Rich clusters of galaxies may contain as many as _____galaxies.
   a. 100
   b. 500
   c. 1000
   d. 3000.

3. The best location for seeing the Magellanic clouds would be
   a. Kitt Peak National Observatory, Arizona
   b. Santiago, Chile
   c. Anchorage, Alaska
   d. Honolulu, Hawaii.

4. The Magellanic clouds are
   a. galaxies
   b. H II regions
   c. spiral arms in the Milky Way
   d. a nearby cluster of galaxies.

5. The Andromeda galaxy and the Milky Way galaxy are both members of
   a. the Pleiades
   b. the Coma cluster
   c. a class known as elliptical galaxies
   d. the Local Group.

6. Why can we not be sure we have discovered all the members of the Local Group?
   a. some may lie behind the Milky Way and be obscured from us
   b. some may be too distant for us to detect
   c. it is not easy to determine if a galaxy is a member of the Local Group, or some other cluster
   d. observations must be carried out from space

7. The stars in the Magellanic Clouds
   a. are similar to stars in our galaxy
   b. contain a lower heavy element abundance than similar stars in our galaxy
   c. are too faint for us to observe
   d. are younger than stars in our galaxy.

8. Spectroscopic observations of iron lines in the gas between galaxies
   a. have never been seen
   b. indicate that the gas from which galaxies formed must have contained some heavy elements
   c. show that galaxies may have a wind akin to a stellar wind which sweeps gas out
   d. are evidence for a previous generation of galaxies similar to Pop. II stars.

9. Superclusters
   a. have as many as 3000 galaxies in them
   b. may be only random concentrations rather than fundamental objects
   c. may have formed when massive gas clouds fragmented
   d. are composed of rich clusters of galaxies.

10. Evidence for massive halos around galaxies includes
    a. the mass obtained by adding all the masses of galaxies in a cluster
    b. the rotation curve derived from spiral galaxies
    c. observations of intracluster gas
    d. all of the above
    e. none of the above.

Answers to Selected Problems From the Text

1. In this problem, we assume that a dwarf elliptical galaxy has an magnitude of about M = −15, and can be detected with an apparent magnitude as faint as m = +20. Such a galaxy therefore could be detected to a distance of:

$$d = 10^{1+\frac{m-M}{5}} \text{ pc} = 10^{1+\frac{20-(-15)}{5}} = 10^8 \text{ pc}$$

Dwarf ellipticals could be detected to a distance of 100 Mpc, which is well beyond the Virgo cluster of galaxies located 15 Mpc away.

3. If the temperature of intracluster gas in a rich cluster of galaxies is 100 million Kelvin, then we can calculate its wavelength of maximum emission from Wien's law:

$$\lambda_{max} = \frac{3 \times 10^7}{T} \text{ Å}$$

$$\Longrightarrow \lambda_{max} = \frac{3 \times 10^7}{10^8} = 0.3 \text{ Å}$$

The maximum emission is in the X-ray portion of the spectrum.

Answers to Self-Test

1–d, 2–c, 3–a, 4–c, 5–b, 6–a, 7–d, 8–b, 9–a, 10–d

Additional Readings

Chincarini, G. and Rood, H. J. 1980, The cosmic tapestry, *Sky and Telescope*, **59**(5):364.

Geller, M. J. 1978, Large-scale structure of the universe, *American Scientist*, **66**:176.

Gorenstein, P. and Tucker, W. 1978, Rich clusters of galaxies, *Scientific American*, **239**(5):98.

Gregory, S. A. and Thompson, L. A. 1982, Superclusters and voids in the distribution of galaxies, *Scientific American*, **246**(3):106.

Gribben, J. 1981, How galaxies form, *Griffith Observer*, **45**(4):2.

Groth, E. J., Peebles, P. J. E., Seldner, M., and Soneira, R. M. 1977, The clustering of galaxies, *Scientific American*, **237**(5):76.

Hausman, M. A. 1979, Galactic cannabalism, *Mercury*, **8**(6):119.

Hirshfeld, A. 1980, Inside dwarf galaxies, *Sky and Telescope*, **59**(4):287.

Hodge, P. W. 1981, The Andromeda galaxy, *Scientific American*, **244**(1):88.

Larson, R. B. 1977, The origin of galaxies, *American Scientist*, **65**:188.

Larson, R. B. 1979, The formation of galaxies, *Mercury*, **8**(3):53.

Meier, D. L. and Sunyaev, R. L. 1979, Primeval galaxies, *Scientific American*, **241**(5):106.

Rees, M. J. and Silk, J. 1970, The origin of galaxies, *Scientific American*, **222**(6):26.

Rubin, V. C. 1973, The dynamics of the Andromeda nebula, *Scientific American*, **228**(6):30.

Rubin, V. C. 1983, Dark matter in spiral galaxies, *Scientific American*, **248**(6):96.

Tully, R. B. 1982, Unscrambling the local supercluster, *Sky and Telescope*, **63**(6):550.

Chapter 30

Universal Expansion and the Cosmic Background

Now that we have discussed the components of the universe, stars and galaxies, we undertake a discussion of some of the overall properties of the universe itself. One obvious question arises: did the universe have a beginning, and can we learn anything about it? In this chapter we discuss the observational evidence for our theory of the evolution of the entire universe.

One of the most important discoveries of this century was made by Hubble in 1929, when he showed that the speed of recession of a galaxy is directly proportional to its distance. This relationship is not apparent for very nearby galaxies, but as more distant ones were observed, it became clear that all showed radial velocities directed away from us. By using various standard candles to estimate the distance to galaxies whose velocities he had measured, Hubble was able to arrive at the conclusion that the universe is expanding. In order to maintain regular spacing during the expansion, the separation of galaxies must increase at a rate proportional to the distance between them. In equation form, Hubble's discovery is written $v = Hd$, where v is the radial velocity of a galaxy, d is its distance (in Mpc), and H is a constant called the Hubble constant. This relation holds, no matter which galaxy you view the universe from, and does not mean that our galaxy is in the center of the universe.

The relationship discovered by Hubble implies that at some finite time in the past, all matter was concentrated at a point. We can find how long ago this occurred from the rate of expansion, for according to the Hubble relation, the age of the universe is $1/H$. The value of H is difficult to determine, but current estimates give a value between 50 and 100 km/sec/Mpc, and therefore an age for the universe of about 18 billion years. Another important implication of the expansion of the universe is that it can be used to determine distances to galaxies for which no standard candles exist.

With the realization that the universe started as a very dense point, scientists began to study what the conditions would have been like then. This "big bang" cosmology predicts that radiation from the initial explosion should still fill the universe, although the cooling due to expansion would have shifted the maximum wavelength far into the microwave region. Independent of the theorists working on this problem, radio astronomers at Bell Laboratories detected the background radiation in 1965. Observations at different wavelengths have confirmed that this is indeed thermal radiation corresponding to a temperature of 2.7 K. As predicted, it is isotropic and shows no clumpiness, although recently the motion of the earth with respect to the background radiation has been measured.

Words and Phrases

anisotropic	isotropic
big bang theory	microwave background
expansion rate of the universe	3° background radiation

Exercise: The Expanding Universe

The Hubble relation is the simplest algebraic equation possible, although the data necessary to establish it were very difficult to obtain. In this exercise you will plot velocity versus distance for a number of clusters of galaxies. Although the relationship of velocity and distance is described as linear (in other, words, plotting one against the other produces a straight line) you will find that the actual data do not lie exactly on a single straight line. The value of H, the Hubble constant, is just the slope of the line drawn through the points in your graph. Since there is an error associated with each measurement, you must decide how the errors affect your determination of H. Errors in the measured velocities are not large. Let us assume them to be correct to ±100 km/sec. Errors in the distance measurements, however, as determined from standard candles, can be quite large, and they increase with distance. Suppose these errors are 10 percent of the distance. Plot the data in the following table and put these error bars on the data points in your figure. Next draw what you consider to be the best straight line through the points, and determine H from the slope of this line (the slope of the line is simply the vertical coordinate at any point on it, divided by the horizontal coordinate for that point).

Now answer the following questions, using your diagram:

1. Use your value of H to calculate the age of the universe. What assumptions can you think of that are implicit in the age you have computed?
2. A cluster of galaxies in Hydra has a velocity of 60,600 km/s. How far away is it, in Mpc? In light years?

Clusters of Galaxies*

Cluster	Distance	Radial Velocity
Virgo	19 Mpc	1,150 km/sec
Pegasus	65	3,800
Pisces	66	5,000
Cancer	80	4,800
Perseus	97	5,400
Coma	113	6,700
Hercules	175	10,300
Ursa Major I	270	15,400
Leo	310	19,500
Gemini	350	23,200
Corona Borealis	350	21,600
Bootes	650	39,400
Ursa Major II	680	41,000

* Adapted from Allen, C. 1973, *Astrophysical Quantities* (London: Athlone Press 1973), p. 292

Self Test

1. The value of Hubble's constant, H, depends on
   a. the rate at which the universe is expanding
   b. the units used to measure distance to galaxies
   c. the units used to measure the velocity of galaxies
   d. all of the above
   e. none of the above.

2. Hubble's Law
    a. indicates that the Milky Way is near the center of the universe
    b. would probably be discovered by astronomers in any galaxy
    c. states that H = vR
    d. was derived from observations of Cepheids in the Andromeda galaxy.

3. Astronomers estimate the age of the universe to be about
    a. 6 thousand years
    b. 4.5 billion years
    c. 15 billion years
    d. a thousand billion years.

4. It is believed the primeval radiation, whose temperature was originally millions of degrees, has been redshifted to
    a. 273 K
    b. 20 K
    c. 3 K
    d. 0.01 K.

5. The value closest to Hubble's constant (as presently known) is
    a. 10 km/sec/Mpc
    b. 30 km/sec/kpc
    c. 75 km/sec/Mpc
    d. 500 km/sec/Mpc.

6. If d is the distance to a galaxy and v is the galaxy's velocity of recession, Hubble's law states that d = Hv, where H is the Hubble constant
    a. correct
    b. wrong: v = Hd
    c. wrong: v = H + d
    d. none of the above.

7. If a galaxy is receding from us at 15,000 km/sec, its distance, from the Hubble relation, is about _____ megaparsecs
    a. 300
    b. 750,000
    c. 500
    d. 15×10^6.

8. About 7 billion years ago, the heavier atoms now found in your body most probably were
   a. in the sun, undergoing fusion
   b. part of the planet earth
   c. part of a star, which would lose at least some of its mass before dying
   d. part of the hot fireball that created the universe.

9. Proof that the Milky Way was only one of billions of galaxies was first given by:
   a. Hubble's discovery of the distance-redshift relation
   b. photographs resolving individual stars in the Andromeda galaxy
   c. the discovery of the microwave radiation at 3° K
   d. none of the above.

10. The 3° radiation is believed to be:
    a. the combined light from distant stars
    b. the combined light from distant galaxies
    c. the remnant of the early days of the universe
    d. the redshifted radiation from quasars.

Answers to Selected Problems from the Text

2. Three galaxies are situated so that the distance from galaxy A to galaxy C is twice the distance from A to B. After a long period of time, the distance between galaxies A and B has doubled. If the expansion is uniform then all distances change by the same factor, so that the distance between galaxies A and C also is doubled.
   The distance between A and C is still twice the distance between A and B. Note, however, that an observer in galaxy A will find that galaxy C is receding *faster* than galaxy B: galaxy C has covered twice as much distance as galaxy B in the same amount of time, and therefore has twice the recession velocity.

3. A supernova remnant has a radius of 10 pc, and its outer portions are expanding at a speed of 1000 $km\ s^{-1}$. To calculate its age, we simply determine how long it has taken for the remnant to expand to its present size:

$$t = \frac{(10 \text{ pc})(3.26 \text{ pc/ly})(9.46 \times 10^{12} \text{ km/ly})}{1000 \text{ km s}^{-1}} = 3.1 \times 10^{11} \text{ s}$$

The supernova occurred 3.1×10^{11} seconds, or about 10,000 years, ago.

5. We find the recession speeds of the galaxies from the Doppler formula:

$$\frac{v}{c} = \frac{\lambda - \lambda_0}{\lambda_0}$$

and their distances from the Hubble law:

$$d = \frac{v}{H}$$

Assuming that H = 75 km/s/Mpc, we obtain the following results:
The singly ionized calcium line is observed at 3936 Å in Galaxy A (rest wavelength 3933 Å). Its recession speed is found to be 230 $km\ s^{-1}$, and its distance is about 3 Mpc.
The singly ionized calcium line is observed at 4028 Å in Galaxy B. Its recession speed is found to be 7200 $km\ s^{-1}$, and its distance is about 96 Mpc.
The singly ionized calcium line is observed at 3942 Å in Galaxy C. Its recession speed is found to be 690 $km\ s^{-1}$, and its distance is about 9 Mpc.

6. When the temperature of the universal background was 4000 K, its wavelength of maximum emission was:

$$\lambda_{max} = \frac{3 \times 10^7}{T}\ \text{Å} = \frac{3 \times 10^7}{4000} = 7500\ \text{Å}$$

Answers to Self Test

1–d, 2–b, 3–c, 4–c, 5–c, 6–b, 7–a, 8–c, 9–b, 10–c

Additional Readings

Barrow, J. D. and Silk, J. 1980, The structure of the early universe, *Scientific American*, **242**(4):118.

Long, K. R. and Mumford, G. S. 1976, A new look at the Hubble diagram, *Sky and Telescope*, **51**(2):83.

Muller, R. A. 1978, The cosmic background radiation and the new aether drift, *Scientific American*, **238**(5):64.

Schramm, D. N. 1974, The age of the elements, *Scientific American*, **230**(1):69.

Webster, A. 1974, The cosmic background radiation, *Scientific American*, **231**(2):26.

Chapter 31

Peculiar Galaxies, Explosive Nuclei, and Quasars

A number of bizarre extragalactic objects have been discovered in recent years that have a bearing on the question of the formation of the universe. We discuss them now: radio galaxies, Seyfert galaxies, quasars and BL Lac objects.

The quasars are the strangest objects in this set, since the others all have some characteristics of normal galaxies. Quasars look like stars through even the largest telescopes, and originally their peculiarities were discovered because they appeared to be stellar sources of radio emission. Once searches for quasars were made in visible wavelengths, it was found that only about 10 percent are radio sources. Spectra of these objects show unprecedented redshifts, which, if due to the Doppler effect, indicate velocities of recession ranging from 16 to 92 percent of the velocity of light. These are much larger than the velocities of even the most distant galaxies. Any further understanding of quasars requires that one understand the cause of the redshifts. If they are cosmological (that is, due to the expansion of the universe), then quasars are much more distant than galaxies. The light we receive from them has therefore travelled for a very long time to reach us, we see them as they appeared in the early history of the universe. If the redshift is due to some other cause, then our knowledge of physics may be too incomplete to explain them.

Current evidence seems to be favoring a cosmological origin for the redshifts. Furthermore, objects like Seyfert galaxies and BL Lac objects apparently are related to the quasars. Some quasars have been found in clusters of galaxies with the same redshift as the quasar. The BL Lac objects are elliptical galaxies with very bright cores. They have smaller redshifts than most quasars, and their cores have many of the characteristics of quasars: variability, brightness and synchrotron radiation. In many respects they appear to be nearby quasars. The Seyfert galaxies, spirals characterized by bright, blue, variable nuclei with emission lines in their spectra, may also be nearby, less energetic quasars. It may be that quasars are very young galaxies, and BL Lac objects and Seyferts are intermediate stages in their evolution.

Even if we accept the hypothesis that quasars are infant galaxies, it is not clear what their source of energy is. Many quasars are variable, with periods as short as a few days. This limits the size of the light-emitting region to a few light-days at most. The luminosities of quasars exceed those of the brightest galaxies by factors of up to 100 or 1000. Their spectra are complex and show both emission and absorption lines of hydrogen and helium and often carbon, nitrogen and oxygen. One suggestion that has been made to explain their energy source is the collision of antimatter with matter, leading to total annihilation. The possibility of multiple supernova explosions has also been suggested, which would help explain the rapid build-up of heavy elements known to have taken place in the early history of most galaxies. A massive black hole might also produce enough energy to power a quasar, and this is probably the most widely-accepted theory at the moment.

Words and Phrases

BL Lac objects
cosmological distances
quasars
radio galaxies
Seyfert galaxies

Exercise: Quasars at Cosmological Distances

As summarized in the text, there is good evidence that the large redshifts in quasar spectra are due to the expansion of the universe, and that these strange objects represent a very early stage in galaxy formation. To see how they compare in distance with galaxies, let us compare their derived distances with those of the galaxies we used to find the Hubble constant in the exercise in Chapter 21. The following table gives a list of quasars. Add them to the plot you constructed in the exercise in Chapter 21. You will first have to calculate the distances to the quasars in the table, using the Hubble law v = Hd and the value H = 55 km/sec/Mpc. You will probably have to extend both axes of the diagram considerably. Use a different symbol in plotting so that galaxies are not confused with quasars.

After you have made your diagram, answer the following questions:

1. Is there any overlap in the distances of galaxies and quasars?
2. In the case of galaxies, the apparent magnitude of the tenth brightest galaxy in a cluster, a standard candle, decreases smoothly with distance. At 66 Mpc the apparent magnitude is m = 13, at 270 Mpc m = 16, and at 650 Mpc m = 18. Describe how the apparent magnitude of the quasars varies with distance. What interpretation can you make about this?
3. What is the absolute magnitude of a quasar at a distance of 1000 Mpc whose apparent magnitude is 15? Compare your answer with the average absolute magnitude of a giant elliptical galaxy, which is about –23.
4. How long ago did the light from the most distant quasar in the table leave its source?

Table of Quasars*

Quasar	Velocity	Apparent Magnitude
3C2	180,000 km/sec	19
3C9	240,000	18
PH957	260,000	17
3C48	77,000	16
PH1377	210,000	16
3C138	150,000	19
3C147	110,000	18
3C273	48,000	13
3C275	110,000	19
3C334	110,000	17
3C351	77,000	15

* Adapted from Allen, C. 1973, *Astrophysical Quantities* (London: Athlone Press), p. 291.

Self Test

1. In the largest telescopes, a quasar looks like
   a. a faint planet
   b. a star
   c. a dwarf galaxy
   d. a planetary nebula.

2. Radio galaxies are so named because
   a. they can only be detected in the radio region
   b. they were first discovered by radio astronomers
   c. their radio emission is very strong compared to that of normal galaxies
   d. all of the above.

3. BL Lac objects
   a. are galaxies with many properties of a quasar
   b. have two jets protruding from each side
   c. have some of the largest known redshifts
   d. are found at the centers of quasars.

4. Quasars appear to be star-like points in the sky. The real size of a quasar can be estimated from
   a. its emission lines
   b. its redshift
   c. the period of its light fluctuations
   d. 21-cm radio emission.

5. A quasar radio signal varies by a factor of five, returning to its original brightness about every two months. Hence the radio-emitting region is no more than _____ in size.
   a. two light weeks
   b. one light month
   c. two light months
   d. unrelated.

6. Assuming their distance to be cosmological, the absolute luminosity of a quasar can be compared with
   a. a supernova
   b. our entire galaxy
   c. the brightest known galaxies
   d. tens or hundreds of bright galaxies

7. Based on the redshifts and directions of quasars, it appears that the universe
   a. is expanding
   b. is contracting
   c. seems to be changing with time
   d. is dying.

8. As we look at quasars with greater and greater redshifts, the wavelength spacing of the emission lines
   a. increases
   b. decreases
   c. remains the same
   d. depends on the value of the Hubble constant.

9. The absorption lines in the spectra of quasars are probably formed by
   a. interstellar gas in the Milky Way
   b. hot gas in the outer layers of quasars
   c. the halos of galaxies between us and the quasars
   d. the interplanetary wind.

10. Which of the following statements concerning radio galaxies is not true?
   a. most radio galaxies are elliptical galaxies
   b. the radio emitting regions tend to be asymmetrically distributed with respect to the core of the galaxy
   c. no one yet knows how the radio emitting material was ejected from these galaxies
   d. none of the above.

Answers to Selected Problems From the Text

1. The red hydrogen line with rest wavelength 6563 Å has a full-width of 200 Å (half-width of 100 Å) in the spectrum of a Seyfert galaxy. The maximum speed of the emitting gas can be calculated from the Doppler formula:

$$\frac{\mathrm{v}}{\mathrm{c}} = \frac{\delta\lambda}{\lambda_0} = \frac{100\text{Å}}{6563\text{Å}} = 0.015\ \mathrm{c} = 4600\ \mathrm{km\ s^{-1}}$$

3. If the brightness of a quasar changes in a time of one day then its diameter must be smaller than one light-day (about 26 billion km). The major axis for

Pluto is about 80 AU (semimajor axis of 40 AU), or about 12 billion km. The diameter of the quasar is only about twice the major axis of Pluto's orbit; the quasar is not much bigger than our solar system, yet shines with the brightness of a trillion Suns (from problem 2 above).

Answers to Self-Test

1–b, 2–c, 3–a, 4–c, 5–c, 6–d, 7–c, 8–a, 9–c, 10–b

Additional Readings

Blandford, R. D., Begelman, M. C., and Rees, M. J. 1982, Cosmic jets, *Scientific American*, **246**(5):124.

Bradt, H. and Margon, B. 1978, In search of X-ray quasars, *Sky and Telescope*, **56**(6):499.

Burbidge, E. M. and Lynds, C. R. 1970, The absorption lines of quasars, *Scientific American*, **223**(6):22.

Chaffee, F. H. 1980, The discovery of a gravitational lens, *Scientific American*, **243**(5):60.

Disney, M. J. and Veron, P. 1977, BL Lac objects, *Scientific American*, **237**(2):32.

Goldsmith, D. 1977, Exploding galaxies, *Mercury*, **6**(1):2.

Gott, J. R. 1983, Gravitational lenses, *American Scientist*, **71**(2):150.

Feigelson, E. O. and Schreier, E. J. 1983, The x-ray jets of Centaurus A and M87, *Sky and Telescope*, **65**(1):6.

Green, L. C. 1977, Radio galaxies and quasars, *Sky and Telescope*, **54**(5):384.

Hamilton, D., Keel, W., and Nixon, J. F. 1978, Variable galactic nuclei, *Sky and Telescope*, **55**(5):372.

Kaufmann, W. 1978, Exploding galaxies and supermassive black holes, *Mercury*, **6**(5):78.

Kellerman, K. I. 1973, Extragalactic radio sources, *Physics Today*, **26**(10):38.

Lawrence, J. 1980, Gravitational lenses and the double quasar, *Mercury*, **9**(3):66.

Margon, B. 1983, The origin of the cosmic x-ray background, *Scientific American*, **248**(1):104.

Meier, D. L. and Sunyaev, R. A. 1979, Primeval galaxies, *Scientific American*, **241**(5):106.

Morrison, P. 1973, Resolving the mystery of the quasars?, *Physics Today*, **26**(3):23.

Osmer, P. S. 1982, Quasars as probes of the distant and early universe, *Scientific American*, **246**(2):126.

Pacini, F. and Rees, M. J. 1973, Rotation in high-energy astrophysics, *Scientific American*, **228**(2):98.

Pasachoff, J. M. 1976, Quasars, *Griffith Observer*, **40**(6):2.

Schmidt, M. and Bello, F. 1971, The evolution of quasars, *Scientific American*, **224**(5):54.

Shields, G. A. 1978, Are black holes really there?, *Astronomy*, **6**(10):6.

Smith, H. 1978, Quasi-stellar objects, *Mercury*, **7**(4):27.

Strom, R. G., Miley, G. K., and Oort, J. 1975, Giant radio galaxies, *Scientific American*, **233**(2):26.

Tenn, J. S. 1979, Cosmic distances and QSO's, *Mercury*, **8**(4):67.

Tucker, W. H. 1977, Astronomy's new superstars: supergiant galaxies, *Griffith Observer*, **41**(11):2.

Wyckoff, S. and Wehinger, P. 1981, Are quasars luminous nuclei of galaxies?, *Sky and Telescope*, **61**(3):200.

Chapter 32

Cosmology: Past, Present and Future of the Universe

The study of cosmology can easily become a philosophical, rather than a scientific, quest. Nevertheless, there are some questions that have objective answers, as long as we make certain clear assumptions at the beginning. In the preceding chapters describing the observational properties of the universe, we have discussed the data available for helping answer fundamental questions; now we discuss the interpretation of these data.

It is usually assumed that the universe is homogeneous and isotropic, a combination of properties known as the Cosmological Principle. These assumptions mean that there is no special place in the universe that is different from other places. To analyze the history, structure, and future of the universe requires a mathematical framework, and this is provided by Einstein's theory of general relativity, which describes, in the form of a set of relations called the field equations, the interaction of matter, energy, and gravitation. The basic tenet of general relativity is that accelerations due to gravity and to motion are equivalent. One implication is that space is in a sense curved in the presence of massive objects, so that photons of light follow curved paths. Since the universe contains matter that produces gravitational acceleration, it has a general curvature, which can be negative (corresponding to the so-called open universe, in which the expansion will continue forever) or positive (corresponding to the closed universe, in which the expansion will ultimately be replaced by contraction).

Current observational evidence favors the open universe. One test is to assess the total matter content of the universe, for this determines the amount of gravitational acceleration. The average density contained in all the forms of matter that can be observed is only at best about 10 percent of the so-called critical density, the density required to close the universe. There still could be enough matter to reach the critical density, if much of the substance of the universe is in some invisible form, and many suggestions for the nature of the "missing mass" have been made. These include intergalactic black holes and neutrinos permeating space (there is no doubt that the universe is filled with neutrinos, but there is considerable uncertainty about whether or not these strange subatomic particles contain any mass).

A second general type of observational test of the fate of the universe is to measure how rapidly the expansion is decelerating. If it is slowing rapidly, then it is eventually going to stop, and the universe is closed. If, on the other hand, the deceleration is relatively slow, then it will be insufficient to bring the expansion to a halt, and the universe is open. To directly measure the deceleration by measuring the rates of expansion now and in the distant past is not easy, because it requires accurate distance measurements of the most faraway galaxies. There are less direct means of inferring the rate of deceleration, however, such as determining the abundances of elements formed in the big bang expansion. The amount of deuterium formed during the early, hot phase of the expansion, for example, depends on how rapidly the universe was expanding then. The present-day quantity of deuterium in the universe implies that the deceleration has been relatively slow, and that the universe is open.

From well-defined laws of physics, it is possible to reconstruct the sequence of events as the universe expanded from an initially highly compressed condition. As the expansion began from a state of infinitely high density and temperature, it began to cool, and first elementary subatomic particles, then simple atomic nuclei formed. Eventually atoms of hydrogen, the most abundant element, formed as protons and electrons came together, and from this time on, the interaction of matter and radiation essentially ceased. The matter formed stars and galaxies in time, while the radiation simply cooled as the universe expanded, reaching its present temperature of about 3 K.

We can speculate about the ultimate fate of the universe. If the universe is indeed open, as the evidence suggests, it will continue to expand and cool, and more and more of its mass will become dormant in the form of dead stars. If, on the other hand, the universe is closed, then it will eventually start to contract, possibly leading to a new expansion stage, as part of a continual cycling process.

Words and Phrases

closed universe
Cosmological Principle
cosmology
cosmogony
critical density
deceleration (universal)
deuterium
flat universe
homogeneous universe
isotropic universe
matter-dominated universe
open universe
quarks
radiation-dominated universe

Exercise: The Curvature of the Universe

There are additional tests for the curvature of the universe that have not been discussed here because they are "armchair" experiments; they can be done in principle, but not in practice. This does not diminish their usefulness in helping us understand the very difficult concept of a curved universe, however.

One simple test of the curvature of the universe is to measure the sum of three angles in a triangle. As you know from plane geometry, this sum should always equal 180° (if you have forgotten this, draw a few triangles of different shape, and measure the three interior angles with a protractor; the sum will always be 180°, within the accuracy of your measurements). Now consider what happens if you draw a triangle on the surface of a sphere. You can try this, using a globe, a grapefruit, or any suitable sphere. Draw triangles of different shapes and sizes, and measure the three interior angles, then answer the following questions.

1. Is your answer different than for the plane triangles?
2. Do you get the same answer for triangles of different sizes on the sphere?
3. Suppose you could measure the angles of a triangle formed by three widely-spaced galaxies. How could you determine the curvature of the universe?
4. Why could this experiment not be done by measuring the angles of a triangle drawn from the moon to widely-separated points on the earth?
5. What do you suppose the result would be if you drew a triangle on a saddle-shaped surface (or any concave surface)?

Self Test

1. The edge of the universe
   a. is a force field
   b. is the edge of the matter in the universe, but space has no edge
   c. does not exist
   d. is the edge of both matter and space.

2. Present data seem to indicate that the universe will
   a. expand forever
   b. eventually stop expanding
   c. eventually collapse
   d. none of the above.

3. From outside, the primeval fireball would have looked small, hot, and dense.
   a. correct
   b. wrong: it would have looked large, hot, and dense
   c. wrong: it would have looked large, cool, and dense
   d. misleading: the fireball filled all space, and could not be viewed from "outside".

4. In discussions of cosmology, the "critical density" refers to
   a. the density of matter at the time of the big bang
   b. the present-day density required for the universe to eventually collapse
   c. the density of the universe derived by summing the masses of all visible galaxies
   d. the density of the universe at the point of maximum expansion.

5. Which of the following was created in the big bang?
   a. oxygen
   b. carbon
   c. plutonium
   d. none of the above.

6. If space is negatively curved, the sum of angles in a triangle will be
   a. 180°
   b. 270°
   c. less than 180°
   d. more than 180°.

7. Which of these theories predicts a balance between expansion and deceleration?
   a. inflationary
   b. oscillating
   c. open
   d. closed.

8. Suppose we lived in a time when the universe was contracting. We would observe
   a. that as we looked out in space we were looking forward in time
   b. new galaxies continuously appearing
   c. the greater the distance of a galaxy the greater its blue shift
   d. all of the above.

9. Our current determination of the density of the universe suggests that
   a. the universe will continue to expand forever
   b. the universe is eventually going to begin to contract
   c. space has a positive curvature
   d. we live in a flat universe.

10. Tests to measure the curvature of space have failed so far because
    a. we cannot measure objects at sufficiently large distances
    b. we do not know how these objects evolve over billions of years
    c. we cannot separate the bending of light by interstellar dust from the bending due to curved space.
    d. all of the above.

Answers to Selected Problems from the Text

1. In this problem we convert the number density of the interstellar medium, given as about 1 hydrogen atom for every 10 cm^3, to a mass density. Assuming the mass of a hydrogen atom is about 1.7×10^{-24} grams, the mass density is:

$$\rho = \frac{1 \text{ atom}}{10 \text{ cm}^3} \times \frac{1.7 \times 10^{-24} \text{ g}}{\text{atom}} = 1.7 \times 10^{25} \text{ g cm}^{-3}$$

According to the text, a Hubble constant of H = 55 km/s/Mpc implies a critical density for the universe of $\rho_{\text{crit}} = 6 \times 10^{-29}$ g/cm^3. The density of the interstellar medium is larger than the critical density by a factor of nearly 3000.

There is little implication to the question of whether the universe is open or closed, however: most of the volume of the universe is *not* occupied by galaxies.

2. In this problem we are asked to determine the average density of the universe assuming that there is approximately one galaxy (of about 100 billion solar masses) for every 10 cubic megaparsecs. To determine the density in grams per cubic centimeter, we must convert the mass from solar units to grams and the volume from Mpc^3 to cm^3:

$$\text{mass} : (10^{11} \text{M}_\odot) \times (2 \times 10^{33} \text{ g/M}_\odot) = 2 \times 10^{44} \text{ grams}$$

$$\text{volume} : (10 \text{ Mpc}^3)(3.1 \times 10^{24} \text{ cm/Mpc})^3 = 3 \times 10^{74} \text{ cm}^3$$

The density, then, is:

$$\rho = \frac{2 \times 10^{44} \text{ g}}{3 \times 10^{74} \text{ cm}^3} = 7 \times 10^{-31} \text{ g cm}^{-3}$$

This is *smaller* than the critical density by a factor of nearly 100 (assuming a critical density of $\rho_{\text{crit}} = 6 \times 10^{-29}$ g/cm^3). Based on visual counts of galaxies, the assumptions made in this problem are fairly realistic; there would have to be a great deal of "dark matter" in order to close the universe.

Answers to Self Test

1–c, 2–a, 3–d, 4–b, 5–d, 6–c, 7–b, 8–c, 9–a, 10–b

Additional Readings

Abell, G. 1978, Cosmology: the origin and evolution of the universe, *Mercury*, **7**(3):45.

Callahan, J. J. 1976, The curvature of space in a finite universe, *Scientific American*, **235**(2):90.

Field, G. B. 1982, Unanswered questions in astronomy, Part 2: the hidden mass in galaxies, *Mercury*, **11**(3):74.

Gaillard, M. K. 1982, Toward a unified picture of elementary particle interactions, *American Scientist*, **70**(5):506.

Geller, M. J. 1978, Large-scale structure in the universe, *American Scientist*, **66**:176.

Gott, J. R., Gunn, J. E., Schramm, D. N., and Tinsley, B. M. 1976, Will the universe expand forever?, *Scientific American*, **234**(3):62.

Harrison, E. R. 1974, Why the sky is dark at night, *Physics Today*, **27**(2):30.

Lawrence, J. 1978, The future history of the universe, *Mercury*, **7**(6):132.

Layzer, D. 1978, Is the universe running down?, *Griffith Observer*, **42**(2):10.

Margon, B. 1975, The missing mass, *Mercury*, **4**(1):2.

Overbye, D. 1977, Out from under the cosmic censor: Stephen Hawking's black holes, *Sky and Telescope*, **543**(2):84.

Page, D. N. and McKee, M. R. 1983, The future of the universe, *Mercury*, **12**(1):17.

Pasachoff, J. M. and Fowler, W. A. 1974, Deuterium in the universe, *Scientific American*, **230**(5):108.

Penzias, A. A. 1978, The riddle of cosmic deuterium, *American Scientist*, **66**:291.

Schramm, D. N. and Wagoner, R. V. 1974, What can deuterium tell us?, *Physics Today*, **27**(12):40.

Silbar, M. L. 1982, Neutrinos: rulers of the universe?, *Griffith Observer*, **46**(1):9.

Silk, J. 1975, The evolution of the universe: a lay person's guide to the big bang, *Griffith Observer*, **39**(7):13.

Tinsley, B. M. 1975, Expanding universe, *Griffith Observer*, **39**(11):2.

Tinsley, B. M. 1977, The cosmological constant and cosmological change, *Physics Today*, **30**(6):32.

Trefil, J. S. 1980, Einstein's theory of general relativity is put to the test, *Smithsonian*, **11**(1):74.

Trefil, J. S. 1983, How the universe began, *Smithsonian*, **14**(2):32.

Trefil, J. S. 1983, How the universe will end, *Smithsonian*, **14**(3):72.
Tucker, W. and Tucker, K. 1982, Exploring the universe: a question of galaxies, *Mercury*, **11**(5):151.
Will, C. M. 1972, Einstein on the firing line, *Physics Today*, **25**(10):23.

Chapter 33
The Chances of Companionship

The possibility of the existence of extraterrestrial life is one that most people find extremely interesting. It is only in recent decades that astronomers, teamed up with biologists, have learned enough to speculate at all meaningfully on this matter. Since we know of no abodes of life other than the earth, there are few data to help with this question, but we believe the solar system was formed naturally along with the sun, and that similar conditions must have arisen many times in the history of the universe. Only the tremendous distances between the stars prevents our exploring them in search of other planets.

Life developed on earth within the first billion years of its formation. At that time, the atmosphere probably consisted mainly of hydrogen, ammonia, methane and water vapor. The action of ultraviolet radiation and lightning may have led to chemical reactions which produced amino acids, the basis of living matter. Amino acids have been found in meteorites, and may also be present in dense interstellar clouds, so these molecules must form fairly easily. From there, DNA and RNA had to form, a step which has not yet been completely carried out in laboratory experiments. Evolution proceeded slowly at first: for a billion years only algae inhabited the earth, but gradually more elaborate plants developed and altered the atmosphere through photosynthesis. Animals developed during the last billion years, and ancestors of modern man appeared in the last four million years. The important lesson here is that intelligent life has been on the scene for only a tiny fraction of the earth's history.

Has life developed elsewhere? In discussing star formation, we have seen that planets should be a natural by-product, especially around single stars. This assumption should be confirmed by the Space Telescope in the late 1980's. We have no information on the probability of life developing on these planets, however. We can only make some assumptions and from them calculate the chances for making contact with other forms of intelligent life. The range of assumptions predict that the present number of technological civilizations in our galaxy is anywhere between one and a billion!

With the more optimistic estimates in mind, astronomers have begun a few tentative searches for interstellar signals from intelligent life. The distances between stars make communication through electromagnetic signals the only reasonable method, and it is very difficult to know what wavelength to choose. Much of the attempted listening so far has been carried out at the 21-cm wavelength of hydrogen emission because radio telescopes are designed for it, and because it was thought that other civilizations might also reason that this was a likely channel for communication. Large-scale projects dedicated to searching for such signals have been proposed. Even with the most optimistic assumptions about the evolution of intelligent life, it would be necessary to search thousands of stars over a broad wavelength region, and this would be a very expensive project. We will probably not have an answer to the question of extraterrestrial life for some time to come.

Words and Phrases

carbon-based life
DNA
Drake equation
Project Cyclops
SETI
silicon-based life
technological civilization
"water hole"

Exercise: The Drake Equation

The idea behind the Drake probability equation for estimating the number of civilizations in our galaxy is that a seemingly impossible question can often be broken down into a series of questions which are easier to answer. Here we will apply the same principle to another problem, to develop a feeling for why this works.

Suppose that you are asked how many home telephones in the U.S. are in use at this moment. Let us estimate the answer. We can break it down into a series of probabilities, as follows:

let N be the number of telephones in use at any given time;
let P be the population of the U.S.;
let T be the number of telephones per person;
let F be the frequency of telephone calls per telephone; and
let L be the length of the average call.

The population of the U.S. times the number of phones per person gives the total number of phones in the U.S. Multiplying this number times the frequency of calls per telephone yields the total number of calls per day in the U.S., and multiplying that by the average length of each call (in units of days, since our factor F is in units of calls per day) gives the number of phones in use at any given moment. Thus, the value of N is simply the product of our four factors:

$$N = P\,T\,F\,L.$$

Let us now estimate the values for the individual terms in this equation. The U.S. population is approximately $P = 2.3 \times 10^8$. Suppose there is one home phone for every four people, so the number of telephones per person is 1/4; i.e. $T = 0.25$. Let us guess that the frequency of calls per telephone is $F = 5$ per day, and that the duration of the average call is 10 minutes, or $L = 0.007$ days. Multiplying all of our factors together to calculate N yields

$$N = P\,T\,F\,L$$

$$= (2.3 \times 10^8)(0.25)(5)(0.007)$$

$$= 2 \times 10^6 \text{ phones in use at any one time.}$$

Now try the following problem. What is the number of freckle-faced, blond, left-handed, 8 year old boys in the U.S.? Use your answer to predict how many live in your town. Does this number seem reasonable?

Self Test

1. Why is Drake's statement of the probability of intelligent life on other planets applied only to the Milky Way galaxy?
   a. most other galaxies are too young to have developed intelligent life yet
   b. other galaxies are deficient in metals, necessary for a technological civilization
   c. we are too unsure of the number of other galaxies
   d. the light travel time is so great that it seems unlikely that we would expect signals from intelligent life in other galaxies.

2. Which of the following is the most important factor in believing that the number of planets with life is very high in our galaxy?
   a. the large number of stars
   b. the large number of planets
   c. the ease in forming organic molecules
   d. the lifetime of a civilization.

3. Suspected planetary systems are most often detected in
   a. spectroscopic binaries
   b. eclipsing binaries
   c. astrometric binaries
   d. visual binaries.

4. A planet orbiting another star could be detected most easily by
   a. its reflected light
   b. its gravitational effect
   c. watching for it to eclipse its parent star
   d. infrared observations.

5. If the history of the earth were compressed into one day, the length of time which humans have lived on the planet would be
   a. two hours
   b. half an hour
   c. one minute
   d. ten seconds.

6. The atmosphere of the earth has contained significant amounts of oxygen for _____ of the earth's history.
   a. all
   b. 75 percent
   c. 50 percent
   d. 25 percent
   e. 10 percent.

7. Laboratory experiments have been successful in creating
   a. molecules
   b. amino acids
   c. DNA
   d. bacteria.

8. The wavelength region around 21 cm has been used to search for intelligent signals because
   a. there are no natural signals in this spectral region
   b. radio telescopes are already equipped to make the search
   c. this is the only region of the spectrum in which signals can travel undisturbed by interstellar absorption
   d. all of the above.

9. The reason that carbon is so important to living organisms is that it
   a. is soluble in water
   b. can form long chains and complex molecules
   c. is abundant
   d. combines with oxygen.

10. Why would we not expect to find intelligent life on a planet with a circular orbit at 4 A.U. from an A0 main sequence star, which has 16 times the solar luminosity?
   a. the central star is too luminous
   b. the central star emits too much ultraviolet light
   c. the central star will live a comparatively short time
   d. the planet's orbit must be elliptical.

Answers to Selected Problems from the Text

1. Let us assume that spores travel through the galaxy with a speed of about 10 $km\ s^{-1}$:
   To reach the nearest star, the spore must travel about 1.3 pc, or about 4×10^{13} km. The time required for the journey would be about 4×10^{12} s, or about 130,000 years.
   To travel from the galactic center to the solar system, the spore must cover 10,000 pc, or about 3×10^{17} km. The trip would require about 3×10^{16} s, or about one billion years. Since the galaxy is at least 10 billion years old, there has been plenty of time for such spores (if they exist) to colonize the galaxy.

4. Alpha Centauri is located about 1.3 pc, or 4.2 light-years, away. If the *I Love Lucy* broadcasts started in 1953, then they reached alpha Centauri in 1957. They will reach Canopus (located at a distance of 30 pc, or 98 light-years) in 2051. If anyone lives on planets around Canopus (and if they enjoy the show), we might expect an answer in the year 2149.

5. We are asked to consider which among the 50 brightest stars (appendix 8) might have planets on which life could develop. The answer, of course, depends on what is meant by "life". If restrict ourselves to the possibility of human-like intelligent life, then we can eliminate any star with a lifetime of less than perhaps a few billion years, since there would not be enough time to evolve intelligence. As evidenced by their spectral types, most of the 50 brightest stars are massive, short-lived stars.
   Other considerations by include multiplicity: stable, ecologically suitable orbits are less likely in multiple star systems. Unfortunately, the table provides little information about companion stars in the listed systems. The best guess we can make would be to pick stars with spectral types similar to the Sun. From the list of the 50 brightest stars we find only one, α Cen A, which meets the criterion. Because α Cen A has a relatively close companion (α Cen B, separation about 40 AU), stable planetary orbits might be considered somewhat unlikely (though certainly not impossible).

Answers to Self Test

1–d, 2–a, 3–c, 4–b, 5–c, 6–d, 7–a–b, 8–b, 9–b, 10–c

Additional Readings

Abt, H. A. 1977, The companions of sun-like stars, *Scientific American*, **236**(4):96.

Barber, V. 1974, Theories of the chemical origin of life on earth, *Mercury*, **3**(5):20.

Ball, J. A., Extraterrestrial intelligence: where is everybody?, *American Scientist*, **68**:656.

Bell, T. E. 1978, The grand analogy: history of the idea of extraterrestrial life, *Griffith Observer*, **42**(8):2.

Bova, B. 1975, The search for intelligence, *Astronomy*, **3**(5):6.

Bracewell, R. 1975, Interstellar messengers, *Mercury*, **4**(2):4.

Dyson, F. 1972, Intelligence in the universe, *Mercury*, **1**(6):9.

Herbig, G. H. 1976, A universe teeming with planetary systems — perhaps, *Mercury*, **5**(2):2.

Horowitz, N. H. 1977, The search for life on Mars, *Scientific American*, **237**(5):52.

Hoyle, F. 1978, Astrochemistry, organic molecules, and the origin of life, *Mercury*, **7**(1):2.

Lawless, J. G., Folsome, C. E., and Kronvold, K. A. 1972, Organic matter in meteorites, *Scientific American*, **226**(6):38.

O'Neill, G. K. 1974, The colonization of space, *Physics Today*, **27**(9):32.

Papagiannis, M. D. 1982, The search for extraterrestrial civilizations — a new approach, *Griffith Observer*, **11**(1):112.

Pellegrino, C. R. and Stoff, J. A. 1979, Organic clues in carbonaceous meteorites, *Sky and Telescope*, **57**(4):330.

Pollard, W. G. 1979, The prevalence of earth–like planets, *American Scientist*, **67**:653.

Sagan, C. and Drake, F. 1975, The search for extraterrestrial intelligence, *Scientific American*, **232**(5):80.

Tipler, F. J. 1982, The most advanced civilization in the galaxy is ours, *Mercury*, **11**(1):5.

Trimble, V. 1977, Cosmology: man's place in the universe, *American Scientist*, **65**:76.

van den Bergh, S. 1975, Are we alone?, *Mercury*, **4**(4):8.

Wetherill, C. and Sullivan, W. T. 1979, Eavesdropping on the earth, *Mercury*, **8**(2):23.